# Supply Chain Management at Warp Speed

Integrating the System from End to End

# Supply Chain Management at Warp Speed

## Integrating the System from End to End

Eli Schragenheim

H. William Dettmer

J. Wayne Patterson

CRC Press
Taylor & Francis Group
Boca Raton   London   New York

CRC Press is an imprint of the
Taylor & Francis Group, an **informa** business
AN AUERBACH BOOK

Auerbach Publications
Taylor & Francis Group
6000 Broken Sound Parkway NW, Suite 300
Boca Raton, FL 33487-2742

---

**Library of Congress Cataloging-in-Publication Data**

---

Schragenheim, Eli.
   Supply chain management at warp speed : integrating the system from end to end / Eli Schragenheim, H. William Dettmer, and J. Wayne Patterson.
     p. cm.
   Includes bibliographical references and index.
   ISBN 978-1-4200-7335-5 (acid-free paper)
   1. Production management. 2. Business logistics. I. Dettmer, H. William. II. Patterson, J. Wayne. III. Title.

TS155.S327 2009
658.5'03--dc22

                                                  2008048074

---

**Visit the Taylor & Francis Web site at**
**http://www.taylorandfrancis.com**

**and the Auerbach Web site at**
**http://www.auerbach-publications.com**

We dedicate this book to our patient and understanding wives:

Rela Schragenheim

Diane Dettmer

Cheryl Patterson

# Contents

# Preface

One could call this book a sequel, but it's neither a new version of *Manufacturing at Warp Speed* nor is it a rewrite of the book. The world continually changes, and production management is no exception. This book continues and broadens the scope of the earlier book's material; however, it updates and refines the former approach to production management. It also complements earlier production management tactics with new ideas. Most importantly, however, it integrates manufacturing, distribution, and raw material management into a single end-to-end supply chain.

Human action naturally tends to conform to existing paradigms. We're already heavily invested in these paradigms, and often we have insufficient capacity (or stamina) to challenge them all. Moreover, not every change is an improvement. New thinking—thinking that challenges existing paradigms—also faces one serious inherent obstacle: it's not clear when traditional thinking should stop resisting and just accept the new paradigms. Once we do challenge existing thinking and settle into a new understanding, should we move on in that new paradigm without further digging? And should we elect to dig deeper and, subsequently, discover some "new wrinkles" in the new paradigm, how easy is it to elicit acceptance of such refinements? It's not unusual to fight the same battle all over again. Moreover, when we embrace a new procedure, one that we may have dedicated a lot of thought to, it's possible that it is *not* the best one. There's a lot of emotional pain in realizing that for years we've done things in a flawed or incomplete way. Then there's the question of how long any new insights will continue to be relevant.

Throughout the 1980s and most of the 1990s, Theory of Constraints (TOC) focused on the optimization of an internal constraint. The rationale for doing so was that if the system constraint was within an organization, its performance would be more easily controlled. Consequently, the main objective became improving internal operations (focusing on the weakest link, subordinating everything else to it). The well-known concept of throughput-per-constraint unit (T/CU) is based on having an internal constraint, but even this invaluable metric creates inertia. For example, T/CU is used to guide decisions even when the weakest system link is not a production constraint. For major decisions that might shift an existing constraint, T/CU is clearly not applicable. Even worse, T/CU is often misused in operations to better "exploit" an internal constraint by delaying orders with lower T/CU, frequently violating the company's commitments to the market. This is a sobering example of how what seems to be a good new insight might actually turn into the next flawed policy. And it happens because inertia discourages questioning whether the assumptions behind the policy are still valid.

In *Manufacturing at Warp Speed,* we proposed the idea that the market is always the major constraint. So even if a company's weakest link is a capacity limitation, that capacity-constrained resource (CCR) must *still* be subordinated to the market. If you lack enough capacity to satisfy all customers, at the very least you are obligated to tell your clients what, exactly, you will commit to providing. And when you can't commit to a customer's request, the ethical thing to do is to advise customers of other options (yes, we *do* mean going to a competitor).

When *Manufacturing at Warp Speed* was published, we suggested that enterprise resource planning (ERP), which is based on an old—and definitely flawed—paradigm, required manipulation to effectively support Simplified Drum–Buffer–Rope (S-DBR). In the intervening years since, we have concluded that this is no longer a valid assumption. Since 2000, dedicated S-DBR and distribution software packages have been introduced. Describing these software applications is beyond the scope of this book, but suffice it to say that effective application of S-DBR isn't constrained by existing ERP packages any longer. At the end of this preface you'll find Internet links to Web sites where additional materials are available, including in-depth examples of existing TOC software modules that support the ideas presented within the book.

New in this book is the subject of using S-DBR for making to stock. *Manufacturing at Warp Speed* was all about making products to order. But how do you manage demand that is based on promises of immediate availability, especially when the deadlines for these demands are shorter than the shortest manufacturing process? This book also introduces a new subset of making to stock that we refer to as "making to availability" (MTA). Properly applied, MTA has the potential to afford a producer a significant competitive marketing advantage.

Another related concept we challenge in this book, one that TOC was originally based on in the 1980s, is that the term *buffer* always means *time*. Time buffers have been used in TOC applications through the 1980s and 1990s, so how do we dare challenge that? But when it comes to actually making to stock, we need to reconsider our traditional concept of buffers.

Making to stock naturally leads us to analyze how we *manage* that stock—or rather, product availability. We describe the TOC approach for managing postproduction distribution in detail. And this naturally leads us to the question of managing raw materials. The net result is a fresh look at the supply chain as an *integrated system*—all the way from producing the base materials to the end consumer. We are convinced that TOC has a profound vision for managing supply chains, and this approach has the potential to overcome major undesirable effects that exist in supply chains today.

This new approach to coordinating production with the market that we introduced in 2000 in *Manufacturing at Warp Speed* has since been adopted by many practitioners of the TOC, including Eliyahu M. Goldratt (Israeli physicist who originated the TOC). So this book should be looked upon as a sequel to *Manufacturing at Warp Speed*. All the ideas inherent in that book are still valid. However, with the passage of time, new insights and perspectives have evolved, and we are sharing them here. The last thing we want to see are people stuck in a new kind of inertia. It required time to overcome the inertia of the cost world and move into the throughput world. Likewise, we'd prefer not to simply stagnate in the first notion of what *throughput world* really means. This book advocates moving forward into the next paradigm.

## Web Sites

- Software: http://www.inherentsimplicity.com/warp-speed
- Other Examples and Case Studies: http://www.goalsys.com/supply_chain_at_warp_speed

# Introduction

In the plethora of books and journal articles that include the words "supply chain" in the title, so many ideas, promises, cautions, and treatments are presented that one might expect little else need be written on the subject. While the interest in supply chains spans only a couple of decades, and we have learned much about the concept during that time, most experts agree that we've really just scratched the surface. The subject seems so straightforward that we don't recognize many of the potential problems associated with implementing and surviving in one or more supply chains. Everyone in the chain should "get it," but successful practice is hindered by the lack of seamlessness.

This book is primarily about supply chains. Many people equate the term with distribution of finished products alone. What many people overlook is that distribution networks are inextricably connected to the manufacturing processes that "fill their pipelines." A true supply chain encompasses every step of activity from the provision of raw materials to the end user.

Thus, it makes no sense to talk about "optimizing supply chains" without also considering manufacturing flow and securing acceptable raw material. Consequently, any book about optimizing supply chains should set the stage with a brief discussion—in this case, a review—of the production process that "pressurizes the pipeline." Similarly, we can't truly optimize manufacturing flow without reliable partners, both upstream and downstream from the manufacturing process. Moreover, we need a means of linking these members of the chain so that they are "all in this together." Securing upstream support for the manufacturer alone doesn't get the finished product to the end user, unless we also have successful downstream distribution.

We've divided this book into four major parts:

**Part 1**: Simplicity and Manufacturing Flow
**Part 2**: Advancements in Make-to-Order (MTO) and Make-to-Stock (MTS)
**Part 3**: Distribution and Raw Material Management
**Part 4**: The Whole Supply Chain Perspective

As you read and think about each part, keep in mind that our discussion is limited in scope to principally Theory of Constraints (TOC)-related applications.

Part 1 is composed of four chapters. The first chapter discusses the need for tools that are inherently simple. This "inherent simplicity" should extend beyond the concept alone and be equally as simple in execution and regular application. By way of review, Chapter 2 provides a synopsis of the traditional TOC production system: drum–buffer–rope (DBR). This discussion includes the basics of the DBR system, as it was originally conceived by Eliyahu Goldratt to facilitate flow on the shop floor.

Chapter 3 summarizes the subsequent evolution of simplified drum–buffer–rope (S-DBR) from the traditional version. It highlights the advantages S-DBR enjoys over traditional DBR and other shop floor control systems. With S-DBR, the focus shifts from solely managing and subordinating to an internal capacity-constrained resource (CCR, or bottleneck) to considering market demand as the primary (ultimate) constraint. This naturally implies a modification to traditional DBR practice. S-DBR steers us toward monitoring the weakest internal link, while simultaneously ensuring enough protective capacity to meet all commitments to the market. *Manufacturing at Warp Speed* (CRC Press, 2000) addressed all of this in detail. None of that changes with this book, so it isn't repeated here. Nevertheless, the technique of tying subordination to the market with the load on the weakest link has significantly improved.

Chapter 4 reinforces the idea that management's challenge is really to create a continuing, careful balance between the internal weakest link and the real demands of the market. This is a highly dynamic balance, and an organization that responds well to real market needs should be able to grow quickly enough to avoid compromising market expectations. Supporting elements of S-DBR (how to estimate safe delivery dates, how safe dates affect other measures, etc.) are also addressed. The new concept of "safe dates" promotes better scheduling, organizing, and synchronizing, fully capitalizing on the planned load concept. Finally, we visit a variety of other issues related to capacity, reliability, and value management that are somewhat unique to S-DBR.

Part 2 begins by addressing whether S-DBR might be too simple to deal with complex situations. Chapter 5 puts forward the idea that simplicity, a principal characteristic of S-DBR, offers a competitive advantage because it is based on (and adds to) the strengths of traditional DBR. Time and again, simple solutions have demonstrated a high probability of working effectively. But at the same time it's natural to question whether S-DBR might not be "just too simple." This chapter examines several production complications and identifies the "tailoring" required for successful S-DBR implementation in nontypical situations.

Chapter 6 addresses the application of TOC tools to MTS environments. Additionally, we introduce the idea of making to availability (MTA), a much more discrete subset of making to stock that affords a potentially powerful competitive advantage. S-DBR, like its traditional predecessor, prompts companies to shift their emphasis from producing for finished inventory to filling firm customer orders. The underlying assumption was that, in many situations, faster production flow makes producing to order feasible. This, in turn, allows significant reductions in finished goods inventories. However, such dramatic reductions in production lead time as S-DBR affords are still not short enough for some customers. Yet no company wants to forego that business. The solution is guaranteeing (for selected customers) product availability. In this chapter, we discuss some of the specific problems, and how to overcome them, in MTA. Some cautions and suggestions for implementation are also provided.

Chapter 7 examines challenges in organizations that include a mixture of make-to-stock, make-to-availability, and make-to-order. Some of these challenges include seasonality, dependent setups, and situations in which there is only one opportunity to "get it right the first time." In making to availability, tighter load control is required. That extra load control must ensure enough protective capacity to offset Murphy's law as well as any sudden shift in priorities.

Part 3 includes two chapters that address the components of the supply chain that are downstream and upstream from manufacturing. Chapter 8 describes the TOC approach to managing *distribution networks*—the distribution portion of supply chains. Manufacturing products alone doesn't assure a sale, even if these products are actually in demand. The choice of having products reach end-users or storing them in warehouses is a real "no-brainer." Without sales to consumers, finished goods inventories are not drawn down. Future production of those products ultimately ends. This is where distribution networks come to the rescue of producers, at least in the case of consumer products. Chapter 8 considers a full range of issues that can be effectively addressed to extend into the distribution part of the supply chain the kind of rapid flow possible in manufacturing.

Chapter 9 addresses the upstream portion of the supply chain. Suppliers of raw materials and components to the manufacturers are critical to successful production and, ultimately, to effective distribution of end products. The most critical raw material management issue is the inconvenience, slowdown, or blockage that production ensues if raw materials are unavailable. In an effort to improve raw materials management, some companies may erroneously strive for cost reductions, rather than focusing on avoiding the potential damage from raw material stockouts. This damage is out of proportion to the assumed cost saving. It includes loss of reputation in the market and indeterminate opportunity costs. Factors essential to building good relationships with suppliers are identified.

Part 4 contains a single chapter. Chapter 10 points out the necessity for supply chain members to consider their fates tied to the success of the overall supply chain, and not just the result of their efforts alone. In other words, the satisfaction of a member-company's goal is, in the long run, directly dependent on the success of the whole supply chain in executing its mission. TOC has always recognized that the effectiveness of a solution resulting in a "win" for one side, but something "less than a win" for the other side, has a very short half-life. Chapter 10 provides a comparison of business-as-usual with partnering principles that can ensure win–win solutions for all members of the supply chain. Because the chain concept is so straightforward, we often overlook the changes in common business practice—especially in management behavior—that may be required to make a supply chain successful. Implementation is complicated by the fact that many firms are members of multiple supply chains. For any supply chain to succeed, the members must recognize that, in the long run, they sink or swim together. This concluding chapter identifies several dilemmas that supply chain members must resolve to ensure win–win outcomes.

By the time you have finished reading this book, you will have been exposed to "the state-of-the-art" in applying the latest TOC methods to manage supply chains effectively. Solutions that represent a "win" for both end users and supply chain members may not be bullet proof. However, we would argue that they have "more Kevlar®"* than other alternatives.

---

* Kevlar is a registered trademark of E. I. du Pont de Nemours and Company.

# The Authors

**Eli Schragenheim** is managing director of Elyakim Management Systems, Ltd., and an international expert in the Theory of Constraints (TOC) and its links to other management philosophies. He is the author of *Management Dilemmas* (CRC/St. Lucie Press, 1998); co-author (with Carol Ptak) of *ERP: Tools Techniques and Applications for Integrating the Supply Chain* (CRC/St. Lucie Press, 1999); co-author (with Bill Dettmer) of *Manufacturing at Warp Speed* (CRC/St. Lucie Press, 2000); and co-author (with Eliyahu M. Goldratt and Carol Ptak) of *Necessary But Not Sufficient* (North River Press, 2001). A former partner of the Avraham Y. Goldratt Institute, Schragenheim developed the original TOC computer simulators used in Constraint Management education. His Management Interactive Case Study Simulator™ (MICSS) is the basis of Goal Systems International's two-day course, "The TOC Challenge," and the three-and-a-half day Simplified Drum–Buffer–Rope (DBR) course, "High Velocity Manufacturing in an ERP/MRP Environment." He is also the creator of the Project Management Simulator™, (PM-Sim) a computer simulation designed to evaluate, test, and compare traditional PERT/CPM and Critical Chain schedules on the same real-world projects. Eli's commitment to teaching by hands-on case study makes learning for students both participative and easier. He holds a Bachelor of Science Degree in mathematics from the Hebrew University of Jerusalem and a Masters Degree in business administration from Tel-Aviv University, both in Israel.

**H. William "Bill" Dettmer** is senior partner at Goal Systems International, a consortium of management professionals with more than 120 years of combined experience. He has twenty-three years' experience in military operations, logistics, strategic planning, operational planning, training, large-scale systems deployment, project management, and contracting. Since 1996, Dettmer has consulted with commercial companies, government agencies, and not-for-profit organizations throughout the U.S., South America, Europe, Japan, South Korea, and Australia. He is author of *Goldratt's Theory of Constraints* (ASQ Quality Press, 1997), *Breaking the Constraints to World-Class Performance* (ASQ Quality Press, 1998), *Strategic Navigation* (ASQ Quality Press, 2003), *Brainpower Networking Using the Crawford Slip Method* (Trafford Publishing, 2003), and *The Logical Thinking Process* (ASQ Quality Press, 2007), as well as co-author (with Schragenheim) of *Manufacturing at Warp Speed* (CRC/St. Lucie Press, 2000).

**J. Wayne Patterson** is a professor emeritus of management at Clemson University.

He received his PhD. in business administration from the University of Arkansas and his BS and MA in business statistics from the University of Alabama. His research and teaching interests are in statistics and operations management with emphasis on the theory of constraints (TOC), quality management and maintenance. His research has appeared in *Decision Sciences, Quality Management Journal, Journal of Production Economics, International Journal of Production Research, European Journal of Operational Research, Journal of Business and Economic Statistics*, and other journals.

He received his training in TOC from the Goldratt Institute's Jonah Program and Management Skills Workshops. He has developed and taught courses that incorporate TOC into the undergraduate curriculum and has also directed a Doctoral Seminar on TOC at Clemson University. In addition he has developed and presented TOC training workshops for practicing managers.

# Chapter 1

# Inherent Simplicity

## Contents

All things being equal, simple solutions are usually best. This bit of advice is thought to originate with Occam's Razor*. In other words, any solution or explanation should not be any more complex than absolutely required—or in the contemporary vernacular: "Keep it simple, stupid!" Simplicity is a positive parameter in hard sciences such as physics. In a class on the history of the physical sciences, one professor expressed his opinion about simplicity this way:

---

* Occam's Razor (sometimes spelled Ockham's Razor). A principle attributed to the fourteenth-century English logician and Franciscan friar William of Ockham, stating that the explanation of any phenomenon should make as few assumptions as possible, eliminating those that make no difference in the observable predictions of the explanatory hypothesis or theory. The principle is often expressed in Latin as the *lex parsimoniae* ("law of parsimony" or "law of succinctness"): "*entia non sunt multiplicanda praeter necessitate*," or "entities should not be multiplied beyond necessity." Often paraphrased as: "All other things being equal, the simplest solution is the best." In other words, when multiple competing theories are equal in other respects, the principle recommends selecting the theory that introduces the fewest assumptions and postulates the fewest entities. It is in this sense that Occam's Razor is usually understood. (http://en.wikipedia.org/wiki/Occam's_Razor)

You know, throughout history the great thinkers searched for the *one and only* formula that would explain the whole world. They came up with something that explained part of the world quite nicely. Then they later found some phenomena that did not fit the formula. So they made the formula more complicated to account for the phenomena. After several iterations of this, the formula became exceedingly complicated. Then someone else (such as Copernicus) came up with a very different simple, straightforward formula that explained all the phenomena. Then more phenomena were discovered that did not match this new formula … And so it continues, even these days.*

## Simplicity and Complexity (Chaos) Theory

Most people have heard of chaos theory, also sometimes referred to as *complexity theory*. The explanations of it are often so deep that most people don't really understand what it means. So at the risk of oversimplifying and leading readers to draw the wrong conclusion, here is the five-minute explanation.

Ilya Prigogene is generally acknowledged as one of the fathers of chaos/complexity theory. His ground-breaking work on the subject in 1967 earned him the Nobel Prize in Physics in 1977 [Osinga, 2007]. Chaos theory hypothesizes that the fate of systems are determined by small factors that become magnified over time. Because these factors are too numerous and too small to know, a system becomes unpredictable. However, Prigogene proposed that the behavior of chaotic systems is not merely random, but rather exhibits a deeper level of patterned order.

An actively imposed change on a system represents a perturbation beyond a boundary of some kind. Though this perturbation may be small, because of the nonlinearity of complex systems it can result in a radical regime change—a kind of *disequilibrium*. The philosophical thrust of chaos theory is that uncertainty can be caused by small changes which, even if anticipated, can result in an unpredictable system. Yet this doesn't mean that system behavior is totally unpredictable. Long-term trends can be distilled within a certain probability, and the range of change can be estimated to some extent. However, the sensitivity to initial conditions in many systems forces a shift from *quantitative* toward *qualitative* analysis. Or to paraphrase statistician W. Edwards Deming, the most important aspects of complex systems may be unmeasured and immeasurable.

What this means to a practitioner of constraint theory in a complex organization is that:

---

* Personal experience recounted by Eli Schragenheim, ca. 1964.

1.  A simpler pattern or order—an *inherent simplicity*—underlies the apparently chaotic, complicated behavior of our systems.
2.  Efforts to quantify the relationship between the behavior of a system and a localized change are likely to be fruitless—and frustrating. The best that may be possible is a qualitative assessment that a given change will ease the system into a desired direction.
3.  Localized change introduces a degree of uncertainty into the behavior of a system, but longer-term trends can be estimated with a reasonable probability.

Constraint theory represents a struggle for the aforementioned inherent simplicity. What's ironic, though, is that with all the simplicity the Theory of Constraints (TOC) offers, we tend to complicate it from time to time because we need to be able to explain reality, which often doesn't seem to be in line with the theory. Early students of TOC often respond that it's just common sense to place primary emphasis on the constraint, yet how often is this simple concept unrecognized? With apologies to Mark Twain, he was correct in musing, "Common sense ain't always that common."

When it comes to management, the task of which is to achieve a goal by enlisting the mutual cooperation of a group of people, the role of simplicity is even more crucial. Why? Because complexity doesn't communicate very well among people, and achieving the common goal requires a means of effective communication. If we have to unify and coordinate efforts, we had better have a very simple process. The greater the complexity, the more opportunities there are for oversights, bad assumptions, and breakdowns.

Another reason for avoiding complexity whenever possible is *sensitivity*. Mathematics uses a term called *sensitivity of the solution*. It refers to the magnitude of change in the result from a small change in one input—in economics, this is referred to as elasticity. In other words, suppose one variable changes just a little. By how much will the result (the solution) of the formula change? If the change in result is large, we say that the output is *very* sensitive to changes in the input. Unless such sensitivity is the objective of our efforts, as in an automobile's power steering system, this might not be a desirable situation.

Here's an extreme example. Suppose you're seeking to improve your health by refining your diet. Let's say that someone provides you a detailed formula of what you have to eat every day to optimize your health. This prescription is supposed to provide only benefits, with no adverse effects at all. It sounds like a good idea, but the existence of a good idea doesn't ensure implementation.

Now let's assume that the formula is *very* sensitive: If you eat as little as one gram more than prescribed of the specific bread, the diet's effects are completely neutralized. In the real world, this kind of "solution" is not much use to us at all. We can't be that precise in what we eat. A diet that isn't so restrictive (sensitive) and admits small deviations without much consequence is much better for most of us.

The point is that complex solutions tend to be *very* sensitive. This is the nature of complexity. The interaction between different variables is such that a small deviation in one variable might be magnified by multiple interactions, causing a very significant deviation in the output. And the more degrees of freedom there are in the variables, the greater the sensitivity is likely to be.

## Complexity on the Manufacturing Floor

In light of this discussion about complexity, let's now consider a typical production shop floor. At the outset, it is a very complex environment. You have materials of varying quality going into operations where the actions and decisions of operators have some impact. Timing is often crucial in certain operations. The way operators do certain jobs, and when they do them, has significant implications for the next operation. Then there is scrap, and clients with requirements and last minute changes, and so forth.

Reacting to this scenario, most manufacturing organizations perform in a more or less "adequate" way out of necessity. They may not be very reliable, but maybe not all that bad either, in spite of the significant complexity of their environment. How do such organizations manage to succeed in the face of this complexity?

The answer is that they naturally—perhaps unconsciously—*simplify* their environment, even though they don't always understand it. For instance, by maintaining a lot of excess capacity, most production operations have made capacity issues nonrelevant. In doing so, they've reduced complexity. They may have a substantial number of people cross-trained in different capabilities. This significantly reduces complexity by making precise scheduling of the manpower irrelevant. Interestingly, as soon as one adopts "Lean,"* much of the "wiggle room" in the system is removed (in the interest of "waste elimination"), and the sensitivities of individual variables have even greater impact (Womack and Jones, 1996).

Constraint theory assumes excess capacity in almost all areas of production operations and that, at most, only the weakest link of the production floor really matters for time planning (i.e., ensuring orders are shipped on time).† In this respect, TOC simplifies the complexity of production operations. The most dangerous trap for practitioners of constraint management is to become involved in the natural complexity of production processes, not understanding that production variables and their interaction are all irrelevant for the overall planning.‡ Instead, the focus should be on basic questions, such as the following:

---

* Lean manufacturing addresses the removal of any waste within the production system. We will discuss this more fully in Chapter 3.
† Theory of Constraints also assumes that employees are willing to do what is right for the company. In some cases, this may be an unwarranted assumption. In all cases, instilling this desire is a leadership function and is beyond the scope of this book.
‡ Schragenheim has said, "When I go into a production facility I don't even tour the plant. I'm afraid it'd confuse me."

1. Can you satisfy all the current demand?
2. If not, which *one* resource blocks you?
3. If one does block you, how could it be more effectively utilized? (By "utilization," we're not talking about the technical meaning. We assume the operators know the difference between keeping "busy" and generating real value.)

These are the really important questions. However, some additional questions are also relevant, even when the direction of the solution is already clear:

1. Will it be required or practical to change the behavior of operators?
2. How interchangeable are the operators in our process?
3. Are there other problematic critical parts (poor quality, unreliable supplier, etc.) that affect overall performance?
4. What other variables could influence the details of the solution (e.g., touch time versus lead time, safety, etc.)?

Clearly the overwhelming advantage of simplicity is that *it is practical to adhere to it in reality*. It's so much better than a complex optimization that, on paper, might be able to achieve a little more, but in reality only seldom achieves more than the simple one. Nothing could be simpler than the fact that an hour lost at the bottleneck is gone forever, yet without realizing this and planning to keep the bottleneck busy, the impact of the lost capacity may be overlooked or unrecognized. As Alex, Lou, Stacey, and Donovan learned from Jonah in the book, *The Goal* [Goldratt and Cox, 1992], it's far better to be approximately right than precisely wrong.

The basic message of having a solution that isn't too sensitive to most deviations is that there is no point in being *precise* about it (Goldratt and Fox, 1987). So for example, the size of the buffer can't be precise, and it's futile to attempt to optimize it. Inventories can be "about right" or definitely wrong. This is a description of reality that is simple enough to understand, even for a board chairman, and will allow us to behave in a way that's "good enough," not only to keep our process afloat successfully, but also to make the working climate less stressful. There is no assured way that is really superior to doing so. Thus, why not make it easier on those who are really instrumental in keeping the system running? In other words, keep it simple.

## Planning versus Execution

Complexity might motivate clever, sophisticated people to invest a lot of time and effort into planning, hoping to achieve the ultimate "optimal solution"—achieving the most that the organization can do at minimum cost. We've already noted that dealing with complexity leads to sensitive solutions that crash in reality and eventually yield mediocre results. One critical measure of whether a plan was really "good enough" is whether the plan was executed as originally designed or whether substantial deviations occurred. If a plan doesn't pass this test (having been implemented almost exactly as planned), then it was either too detailed, the impact of uncertainty (Murphy's law*) was not adequately considered, or a major basic assumption underlying the plan was invalid.

Planning means *making decisions ahead of time*. In an uncertain environment, making decisions ahead of time opens the door for "Murphy" to step in. So regardless of the objectives and benefits of planning, each planning decision must be justifiable. Realizing this, we should acknowledge some rules behind effective planning:

- The objective of effective planning is to synchronize a variety of efforts to achieve the desired high-level outcomes of the larger organization. Effective synchronization is essential to the success or the failure of most plans. It requires significant thinking in advance of the activities intended to execute the plan.
- Even what seem to be very complex organizations have a significant amount of excess capacity. This excess capacity reduces the impact of close dependencies between many variables, thus making the environment simpler. This means that an operating environment will have a few areas that are truly critical to achieving results, while many others can adapt quickly to whatever effort is required. As a result, their synchronization can be attained on very short notice.
  - The unavoidable conclusion is that planning should concentrate on the essential synchronization of the really critical areas—those areas where any deviation would compromise objective attainment. Therefore, to be effective, planning must focus on the few system components for which required actions would not tolerate even relatively small deviations.
- Once the truly critical elements of a plan are articulated and the resulting required actions are determined, those critical actions must be protected from operational variation and uncertainty. In short, those actions must be executed as specified. In Theory of Constraints terminology, *buffers* must be included in the plan for protection from uncertainty to which the environment is subjected.
  - In reality, most plans do allow for some kind of buffer. However, in most cases such buffers are embedded in the plan in such a way that even the planners themselves are not conscious of them. For instance, a director of engineering might claim that the development of a certain product feature will take six months. Really? Doesn't such an assertion really mean: "We can probably do it in three to four months, but to be on the safe side,

---

* "Whatever *can* go wrong, *will* go wrong, and it will do so when least expected and most inconvient."

we won't commit to doing it in less than six?" The same phenomenon occurs in budgeting. All departments typically ask for more than they actually estimate they'll need.

Let's use a simple example. You are planning a vacation trip to a destination you've thoroughly researched but never before visited. Would you plan your agenda for each day in great detail? If the weather at your proposed destination isn't terribly predictable, you could find your schedule completely invalidated. Some attractions might be open, others closed. And do you really want to determine ahead of time how long you'll spend at each attraction?

On the other hand, there are some actions you should establish ahead of time. Assuming your vacation time is limited (you must return on a fixed date), you might need to reserve seats on a return flight. Departure for your return, a particularly critical element of your plan, requires a time buffer to ensure that you arrive at the airport on time. This might suggest completing your vacation trip somewhere close to the airport. Similarly, your outbound flight to your vacation destination should likewise be well protected, or the precious time you've planned for your vacation could be needlessly wasted.

The general rule about planning can be summarized this way: *focus on the critical areas and include buffers where required.*

## The Impact of Effective Planning on Execution

Applying this general rule yields several benefits. First, execution becomes much more flexible. The few critical activities or functions are adequately protected from variability and uncertainty by buffers. The more numerous less-critical areas are protected by their excess capacity. Buffers of critical areas can be monitored and excessive buffer consumption mitigated before it compromises the functions that must be strictly performed as specified.

Second, variation and uncertainty are accommodated without adverse impact on the whole organization. In other words, each corrective action to reestablish proper synchronization will have only local ramifications, not broader system impact. For instance, overtime might be required to ensure the appropriate delivery of an order, but the overtime does not result in a global change in the budget. The overall budget remains the same; other budget lines aren't impacted by the overtime used.

Third, people have very clear priorities for successful execution of critical decisions identified in the plan. The whole point of synchronizing the less critical parts of the system is to reduce the operational risk (and effects) from variability. Buffers provide the maneuvering room in which execution decisions can be made. In other words, priorities should be based on the state of the buffers. This realization—the backbone of TOC that we refer to as buffer management—has huge ramifications. The traditional techniques of TOC have only begun to reveal these ramifications.

The relationship between planning and execution is at the very heart of implementing simple, effective management in all complex organizations. Any simplification to the planning must have a corresponding counterpart in execution.

Let's clarify something else. Including visible buffers in a plan makes the plan less obviously "optimal." Buffers testify to the extent of a resource's capacity (money is a resource, too) that might be required. This doesn't mean that such capacity will always be required, but in extreme circumstances, it might be. Consequently, there will always be "sophisticated" people who will try to reduce the buffers, usually in the interest of "efficiency." Or they may try to find the secret of a truly "optimal" buffer. They might as well play with a bomb. In arbitrarily reducing buffers, luck will eventually run out and an "explosion" will happen. We will say this now, and throughout the rest of this book: *Buffers are based on intuitive judgment and are subject to feedback from reality, but never fool yourself into thinking you can ever know the "precisely right-sized" buffer.*

Theory of Constraints methods, both for production and for project management, are based on defining the critical elements that must be planned and buffered, then guiding successful execution to achieve all the planned objectives in spite of "Murphy." Please keep in mind this clear distinction between planning and execution and the way they're synchronized. They are important not only to successful production planning and execution, but to so many other areas as well.

The rest of this book will be predicated on this important principle, so succinctly articulated as Occam's Razor: simpler is better.

# References

Goldratt, E.M. and Jeff Cox, *The Goal: A Process of Ongoing Improvement*. Great Barrington, MA: The North River Press, 1992.

Goldratt, E.M. and Robert E. Fox, *The Race*. Great Barrington, MA: The North River Press, 1987.

Osinga, Frans P.B., *Science, Strategy, and War: The Strategic Theory of John Boyd*. New York: Routledge, 2007.

Womack, James P. and Daniel T. Jones, *Lean Thinking: Banish Waster and Create Wealth in Your Corporation*. New York: Simon & Schuster, 1996.

*Chapter 2*

# Understanding Traditional Drum–Buffer–Rope and Buffer Management

## Contents

Many of the shop floor management control systems used during the late 1970s and early 1980s were somewhat convoluted and often placed their focus on the "wrong" incentive. Conventional wisdom of the time placed a high premium on the efficiency of each step in the manufacturing process. Every workstation had to be kept busy all the time, or it was not considered "efficient." Two such systems bear quick mention before we discuss traditional drum–buffer–rope (DBR). First, large manufacturing operations use computer programs that do *materials requirements planning* (MRP). Many of the MRP implementations also frequently release extra work onto the shop floor to keep work centers in the production process as close to fully occupied as possible [see Ptak and Schragenheim, 2000 for more detail]. As the production planners introduce a large amount of material—work-in-process— into the production floor, the production facilities began to look at lot like Figure 2.1.

MRP originally focused on the master schedule from which the material requirements were generated, but it quickly evolved into a more complex system, encompassing additional factors in resource planning, rough-cut and detailed

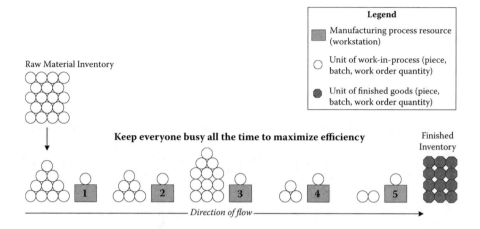

**Figure 2.1  Traditional manufacturing operation (MRP).**

capacity planning, and shop floor control. The natural evolution of MRP to MRP II (an expansion of MRP to include sales, operations, and financial planning) and then to enterprise resource planning (ERP) has expanded the focus beyond that of facilitating the shop floor, leading to a system that includes virtually all aspects of the organization, but even today MRP is still the predominant control system.

Suffice it to say that the firms embracing the ERP packages have agreed to significant expenditures for the computing hardware and software to drive such a system, and the disappointments are often more common than successful implementations [Kumar, Maheshwari, and Kumar, 2003]. ERP systems have been applied to the management of every operation of an enterprise's value chain to minimize costs and time required to deliver products to customers [Xue et al., 2005]. However, while ERP appears well suited to the concept of supply chain management, the difficulty of achieving the elusive benefits suggests that a more simplified approach might be a better option.

Second, a "new" production approach, which experienced the highest visibility in the 1980s, was a creation of the Japanese: *just-in-time* (JIT). An element of the highly successful Toyota Production System, JIT captured the imagination of western manufacturers for two basic reasons. The first was that W. Edwards Deming and Joseph Juran brought the message of Japanese quality to the Western Hemisphere in a highly visible way. Their message had such impact that in industrial circles, Japanese ways of management were often assumed to be better than any other ways. The second, and perhaps even more compelling, reason was that Japanese products, especially automobiles, were soundly beating American products in the marketplace. And again, typical of business culture all over the world, "anything Japanese" was thought not only good, but worth emulating. This included the renowned (and to American manufacturers, scary) Toyota Production System, a key component of which was JIT.

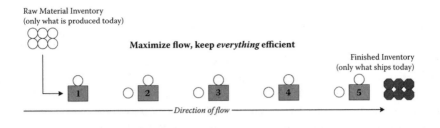

**Figure 2.2    Just-in-time (JIT).**

JIT eliminated almost all inventories in the system: raw material, work-in-process, and finished. Only materials to be worked on during a given day were present in the plant, and what was finished would be shipped the same day or the next (Figure 2.2).

Many manufacturers in the West attempted JIT, but most continued traditional MRP mass production, just as they had been doing for years.* Those manufacturers who continued "business as usual" discovered that they were at a great disadvantage when the quarterly reports came in. Those who adopted JIT discovered that it wasn't quite as easy as the Japanese made it look. For a variety of reasons—some cultural, some quality-related, and some owing to the logistical circumstances—they were never quite able to replicate the stunning results of the Toyota Production System.

In its relentless pursuit of efficiency, JIT took away the safety net:

> To get continuous-flow systems to flow for more than a minute or two at a time, every machine and every worker must be completely "capable." That is, they must *always* be in proper condition to run precisely when needed and every part must be made exactly right. By design, flow systems have an *everything-works-or-nothing-works quality*, which must be respected and anticipated (emphasis added) [Womack and Jones, 1996].

The Japanese operated their entire production system—engineering and supply chain as well as production—in a way that ensured the integrity of production. Most Western companies merely took just-in-time out of context and eventually wondered why they couldn't do as well as the Japanese.

A system that came to be known as the Theory of Constraints (TOC) was developing at about the same time as JIT was being first absorbed by Western manufacturers. Within the TOC tool kit, the shop floor control system, called drum–buffer–rope (DBR), is one of its most significant elements. Created by E. M. Goldratt to resolve a "bottleneck" in a job shop manufacturing facility, its application grew in both numbers and scale over the next two decades [Goldratt and Cox,

---

* As Winston Churchill once observed, "Man will occasionally stumble over the truth, but most of the time he will pick himself up and continue on."

2004]. Prior to this time, and still quite common today, management thinking had zeroed in on balancing capacity and merging it with demand using a push system to move work-in-process through the stages of manufacturing. The success of DBR and other variations on a pull-oriented system to replace the "push" style made it clear that there was "a better mousetrap."* The original DBR concept, alluded to in Goldratt's first book, *The Goal* [Goldratt and Cox, 1984], and described in more detail in the subsequent revised editions of *The Goal* [Goldratt and Cox, 1986], [Goldratt and Cox, 1995], and [Goldratt and Cox, 2004]; *The Race* [Goldratt and Fox, 1987] and *The Haystack Syndrome* [Goldratt, 1990], was intended to manage the *flow* of work through a production process rather than managing the capacity. The objective was simple:

> Maximize continuous flow of the products that ultimately produce revenue for the company, while simultaneously minimizing the amount of partially finished work (work-in-process) resident within the production process.

In other words, keep the *rate of flow* of work-in-process as high as the least-capable resource (the bottleneck) can tolerate without creating a "logjam," while simultaneously reducing the *amount* of work-in-process on the production floor to an absolute minimum without starving that same bottleneck. Another way to say this is: *balance the flow, not the capacity*. This requires maintaining the amount of work-in-process on the production floor within some fairly narrow minimum–maximum parameters.

## Drum–Buffer–Rope

Traditional DBR is structured to provide the benefits of just-in-time (rapid, continuous flow with minimum inventory), but also to protect against the special cause variation, which normally can't be controlled, and the general cause variation that a company might not find cost effective to eliminate. In other words, unlike just-in-time, it doesn't absolutely depend on *everything works or nothing works*.

**Note:** It must be emphasized that product and process quality cannot be ignored in DBR. A production process that incurs substantial scrap, and especially rework, must have its quality addressed before DBR can deliver its best benefits.† DBR is designed to protect against general cause variation that can't be removed from the system and some (but not all) special cause variation. It does not relieve management of the requirement to operate a robust quality

---

* "Build a better mousetrap, and the world will beat a path to your door."
† At a conference in the 1990s, E. M. Goldratt acknowledged that effective application of DBR presumed that quality was largely under control.

improvement program. In the absence of robust quality, DBR may provide some marginal improvements to a production process, but its promise can never be fully realized.

The DBR system begins by pacing production at the rate allowed by the most constraining component of the process. The schedule for this element—the bottleneck or capacity constrained resource (CCR)—is referred to as the *drum* because it sets the fastest sustainable rhythm that the production system could maintain under the best of conditions—the lowest level of variability or disruption. Clearly, DBR represented a rather radical change in the way manufacturing was done. To say that efficiency was critical at only one work center—the CCR—and not particularly important throughout the rest of the plant was tantamount to heresy. To say that idle time at most work centers was not only acceptable, but even *desirable* was even worse. Yet this is what DBR implied: The speed of workflow through the production floor and the reduction of work-in-process inventory to the minimum consistent with reliable promise dates was more important than local efficiencies.

As a practical matter, the likelihood of the lowest level of variation being a reality is generally accepted to be small indeed. That being the case, protection of the drum schedule required some way to ensure that the schedule could be met even in the face of random variations or disruptions to the process. The *buffer* is the protective device used to ensure that a CCR is not starved and its capacity lost. It represents the amount of time *in advance of need* that work-in-process is scheduled to arrive at a particular control point. The size of a time buffer is not directly related to a work center's capacity. Instead, it's related to the *transfer* of work within the process. Time buffers are a management artifice specifically designed to help production managers retain control of on-time deliveries. For all practical purposes, the time buffer is a function of total uncertainty in the system, including human errors, the effect of Murphy's law, fluctuations in demand, and their impact on the wait time. The size of the time buffer can be adjusted relatively easily if needed.

**Note:** Traditional DBR recognizes three different control points in the production process: the CCR, certain assembly points (downstream of the CCR), and the shipping dock. For the sake of clarity, we will examine mainly the rationale for the CCR's time buffer here. Detailed analysis of time buffers for other control points may be found in references at the end of this chapter. In addition, capacity and stock buffers will be discussed in later chapters.

Once the size of the time buffer is established, we can use it to determine the length of the *rope*. The rope is essentially the action-trigger device. It signals the very first step in production—the "gating" operation—to adjust its rate of introducing raw material into the process. The rate to which the gating operation "meters" the entry of work-in-process is equated to the actual production pace of the CCR. If the CCR suffers slowdowns or interruptions in its activity, the entry of material into the production process is equivalently slowed or stopped.

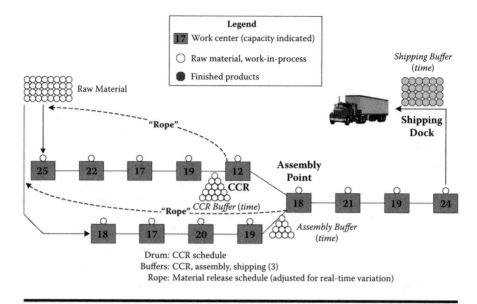

**Figure 2.3  Traditional drum–buffer–rope (DBR).**

This prevents unneeded excess work-in-process from flooding the manufacturing floor. The downside of this result for traditionally minded managers is that nonbottleneck resources will necessarily be idle and, of course, their efficiencies will suffer. Figure 2.3 depicts a typical production process with traditional DBR elements (drum, rope, and buffers) indicated.

## Buffer Management

The plan presented in Figure 2.3 is designed to make best use of the CCR and to ensure the on-time delivery of customer orders, but it's just a plan. How do we ensure that the plan will be executed properly? In reality, there are deviations.* The active discharge of the plan requires a vigilant management scheme—buffer management.

Buffer management divides the time buffer into three equal parts, each of them denotes a different category of priority. During the first part, if the order hasn't reached the CCR, we have no reason to worry. We wouldn't even expect it to be there. This first part of the buffer is the green zone. However, by the time we are into the second part of the buffer, even though we still have ample time, we should at least validate that the materials for the order were actually released. The rule

---

* "The best-laid plans of mice and men often go awry."

is that by the second part of the buffer, production managers might just check where the order is within the shop floor, but not take any real action to expedite. This is the yellow or "monitoring" zone. However, if the final part of the buffer time arrives, but the order has still not shown up at the CCR, the manager should behave very differently. The order is now in the red zone, and immediate action is called for: get that order to the CCR, no matter what. There is just enough time to get it to the CCR on time and still preserve the promised delivery date.

Buffer management does not define the buffers. This is done during production planning. It does, however, monitor every order according to where (in which zone) the particular order resides. Generally speaking the red part of the buffer—the most urgent part of it—prompts production managers to assign a very high priority to the order so that with extraordinary effort (expediting) it might still be delivered on time.

What enables us to complete an order within the red zone is that in manufacturing the net "touch time"—actual processing time along the longest chain of operations—is very small relative to the actual production lead time, certainly less than 10 percent. This is strictly a characteristic of manufacturing; in projects this is not the case. So when a time buffer of 15 days is established, the assumption is that net processing time is not more than 1-2 days. In Chapter 5 we will examine what happens when net processing time is really long.

For example, let's assume that a certain order has a CCR buffer of 15 days. These 15 days are divided into three parts. The first five days following the scheduled release of materials are defined as "green." We expect that it would be perfectly possible to complete the order within the green part of the buffer. This might not be too frequent, but when everything goes well it does happen. The remaining 10 days encompass the other two zones. The zone that begins at day 6 and goes through day 10 is considered the "yellow" zone. The last five, days 11 through 15, constitute the "red" zone (Figure 2.4). When an order is finally completed, it is deleted from the master list of orders. An order that takes more than 15 days, meaning that it's late, is considered "black."

When the time buffer is set at a reasonable value, the majority of the orders will reach the CCR within the yellow zone. Though net processing time may be small relative to the entire 15-day buffer, for some amount of time the order must wait for the required processing resources to become available (that is, to finish other work).

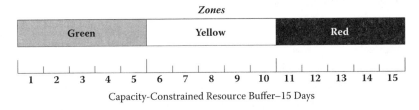

**Figure 2.4   Typical capacity-constrained resource (CCR) buffer.**

This wait time is usually highly variable, unless operators slow the pace of work to keep a continual queue of work waiting for them. However, assuming we succeed in modifying the working culture, we can expect that most orders will require substantially less than 15 days for completion of processing all operations prior to the CCR.

By establishing a 15-day time buffer, we're assuming that it might actually take that long to move the order to the CCR in some cases, but in most circumstances we expect it to be done in less time than that. Establishing this buffer creates an opportunity for orders that may have experienced excessive delay—perhaps the entire yellow zone has been consumed—to be afforded more priority and managerial attention. Clever use of the remaining third of the time buffer—the red zone—can allow the order to finish and ship on time.

To determine whether an order is in danger of late delivery, we use a simple mathematical mechanism. We determine the time consumed so far and compare it with the time buffer assigned. The result provides a straightforward measure of the buffer status—the percentage of *buffer penetration*.

By way of an example, let's continue with the situation illustrated in Figure 2.4 but apply a 15-day buffer to protect the due date of an order even when no CCR is active. Suppose the order has been on the production floor for eight days. Assuming the material release was done exactly 15 days before the promised delivery date, penetration into the buffer—the buffer status, expressed as a percentage—is: $8 \times 100/15 = 53.33\%$. We know that the green zone represents the first 33.33 percent of the entire buffer, and the red zone represents the last 33.33 percent. That makes this particular order "yellow."

Now suppose that the actual material release occurred three days later than scheduled. This is not an unrealistic situation. Let's also say that the order takes the same eight days we cited above to reach the point where it now resides. How much has the buffer been penetrated now? Remember, the date we must account for is the delivery due date. We're now only four days from that date—15 days minus the 3 days for late material release, minus another 8 days for processing thus far (Figure 2.5). Could we consider this order "almost late"?

Calculating the penetration into the buffer now yields: $(15 - 4) \times 100/15 = 73.33\%$. The order is definitely "in the red," as Figure 2.5 indicates. If we expedite it now, rather than waiting until the last day, it will almost certainly still ship on time.

The generic formula for calculating the buffer status (as a percentage) is:

(Time buffer minus Time remaining) × 100/Time buffer

Here is another example. Suppose the order was released three days late, as described above, then it spends four more days on the production floor. It's still in production when the client calls to plead with us for an early delivery—three days early. We could undoubtedly create substantial goodwill with the client if we could satisfy that request. But how would this request affect our production process (Figure 2.6)?

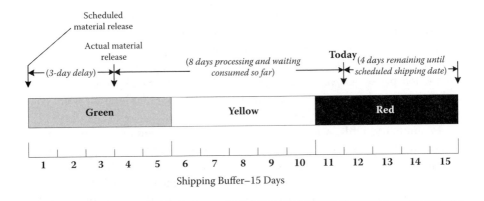

**Figure 2.5    Red zone penetration.**

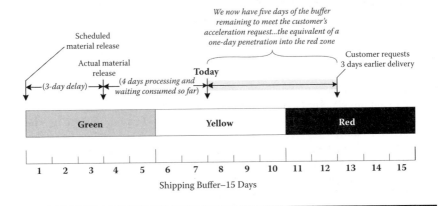

**Figure 2.6    Can we accelerate an order?**

In this situation, the 15-day time buffer is consumed in three different ways: late material release, actual production (and waiting in queue) time, and the customer's request for an accelerated delivery. The relevant question we must answer is: "How much of the buffer has been consumed so far"? The answer is: "Ten days." (Three days late, plus four days production time so far, plus the loss of three days at the back end resulting from the early delivery request.) This means that two-thirds of our 15-day buffer has already been consumed. Will the remaining one-third be enough? If we give it special priority (expediting), it should be. After all, our situation is now the same as if the order had just penetrated the red zone. As long as no other extraordinary factors intervene, we should be able to satisfy the customer's request. (Of course, it would be a good idea to convey to the customer the impression that we've moved heaven and earth to make them happy, whether or not we actually did. Nothing instills customer loyalty like knowing that a supplier will "do the impossible" to satisfy them.)

One more point must be emphasized. Buffer management is *not* concerned with the precise location of the order on the production floor, unless it turns out to be in the red zone. Because the net processing time is relatively short compared with the size of the buffer, the exact location of the order is normally not a major issue. Production personnel *must* know precisely where each *red zone* order is located because these orders require expediting. However, work order location is not considered in calculating the buffer status.

In summary, DBR, a *planning* methodology, goes hand-in-hand with buffer management, an *execution* procedure to help guarantee successful attainment of planned deliveries. Each one is incomplete without the other. In contrast, the blueprint of the DBR system for a particular process is but the first step and the successful implementation of the plan will ultimately depend on the application of buffer management. DBR represents an important step in simplifying shop floor control systems. In the next chapter, we will highlight some of the shortcomings of traditional DBR that led to the "simplified" version (S-DBR) presented in *Manufacturing at Warp Speed* in 2000 [Schragenheim and Dettmer, 2000].

# References

Goldratt, E.M., *The Haystack Syndrome: Sifting Information Out of the Data Ocean*. Croton-on-Hudson, NY: The North River Press, 1990.

Goldratt, E.M. and Jeff Cox, *The Goal: Excellence in Manufacturing*. Croton-on-Hudson, NY: The North River Press, 1984.

Goldratt, E.M. and Jeff Cox, *The Goal: A Process of Ongoing Improvement* (Rev. ed.). Great Barrington, MA: The North River Press, 1986.

Goldratt, E.M. and Jeff Cox, *The Goal: A Process of Ongoing Improvement* (2nd rev. ed.). Great Barrington, MA: The North River Press, 1995.

Goldratt, E.M. and Jeff Cox, *The Goal: A Process of Ongoing Improvement* (3rd rev. ed.). Great Barrington, MA: The North River Press, 2004.

Goldratt, E.M. and Robert E. Fox, *The Race*. Croton-on-Hudson, NY: The North River Press, 1987.

Kumar, V., B. Maheshwari, and U. Kumar, October 2003. An investigation of critical management issues in ERP implementation: Empirical evidence from Canadian organizations, *Technovation*, 23(10): 793–807.

Ptak, C.A., and E. Schragenheim, *ERP Tools, Techniques, and Applications for Integrating the Supply Chain*. Boca Raton, FL: CRC Press LLC, 2000.

Schragenheim, Eli and H. William Dettmer, *Manufacturing at Warp Speed: Optimizing Supply Chain Financial Performance*. Boca Raton, FL: St. Lucie Press, 2000.

Womack, James P. and Daniel T. Jones, *Lean Thinking: Banish Waste and Create Wealth in Your Corporation*. New York: Simon and Schuster, 1996.

Xue, Y., H. Laing, W.R. Boulton, and C.A. Snyder, September 2005. ERP implementation failures in China: Case studies with implications for ERP vendors, *International Journal of Production Economics*, 97(3): 279–295.

# Chapter 3

## Simplified Drum–Buffer–Rope: *An Overview*

## Contents

Drum–buffer–rope's (DBR's) development represented a quantum improvement over materials requirement planning II (MRP-II) as a means of scheduling production operations. Combined with effective buffer management, DBR increased the flow of production operations, decreased the amount of work in process (WIP) in the system, and shortened manufacturing cycle times enough to enable making-to-order in most cases. DBR as described in the previous chapter represents a simplified approach to shop floor control over earlier as well as alternative systems.

In comparing simplified drum–buffer–rope (S-DBR) to DBR, we must consider why S-DBR is indeed even simpler. We still need a "drum" that in S-DBR comes mainly from the market, a "buffer" for protecting the drum, and a "rope" to signal the need to release materials for new orders. So what makes it "simple"? The elimination of two types of buffers (capacity-constrained resource—CCR—and assembly) and the use of the market-driven master production schedule rather than CCR (or drum) schedule clearly simplify shop floor control.

## Rethinking DBR

Traditional DBR was designed for situations where demand was frequently assured, and often even exceeded, a company's capacity to produce. S-DBR recognizes that for a substantial number of companies, during at least part of the time, demand does not consume all of production's capacity. This assumption is reflected in Figure 3.1. Once the framework of S-DBR is in place, it also accommodates the occasional times when a real bottleneck emerges. Later in the book, we will explain

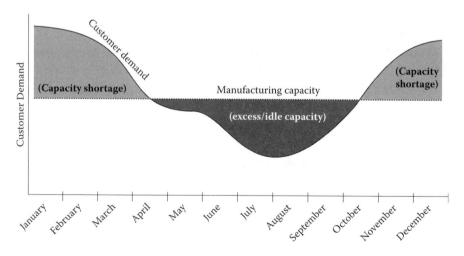

**Figure 3.1   Variations in demand over time.**

in detail some of the recent refinements that enable S-DBR to accurately and effectively smooth the manufacturing load even when the pressure of customer demand is very high.

Some companies experience seasonal peak loads. Other companies may see loads change on a less predictable basis. Either way, during periods of peak demand, manufacturing capacity to produce may fall short of opportunities to sell products. Conversely, during less active periods, demand may not consume all the available production capacity. Most managers consider matching demand as closely as possible to capacity to be a worthy challenge. In the past, some companies have done this by trimming capacity to match demand.* Usually this takes the form of personnel layoffs, which remain in effect until demand begins to surge again. Equipment is merely idled because it's not practical to shed or add equipment repeatedly as demand changes.

## Shipping: The Primary Constraint

S-DBR assumes that customer demand is *always* a constraint to any manufacturing system, but that it is often masked by short-term demand peaks that temporarily drive the constraint inside the manufacturing operation, as shown in Figure 3.1. If this is a valid assumption, and there is ample evidence to demonstrate that this is closer to reality than the original assumption of traditional DBR (demand often exceeds capacity), then effective application of DBR— actually, the entire operations and sales strategy of the company—requires some reorganization.

For one thing, it makes more sense to tie the manufacturing "rope" to actual market demand rather than to utilization considerations. The only alternative is releasing material according to a demand forecast, and forecasts pose inherent problems of their own. Chief among these is their deterioration in accuracy the farther out into the future one projects the forecast. Next week's forecast approaches 100 percent certainty. The forecast for four weeks from now is probably less accurate, and a monthly forecast six months from now carries an associated forecasting error that is very large.

## Scheduling: Based on Firm Orders

Forecasts are obviously necessary for rough-cut capacity planning, but producing to quarterly, semiannual, or annual forecasts for many companies is risky—maybe even foolhardy. How much better it would be to set capacity based on long-range forecasts, but schedule production based on real-time demand. Later, in Chapter 6,

---

* "The glass is neither half full nor half empty—we have twice as much glass as we need!"

we will see how some new ideas about S-DBR address situations in which producing to stock is absolutely necessary while still striving to produce as closely to real demand as possible.

Another concept that must be seriously reconsidered when customer demand is always assumed to be a system constraint is the importance of internal schedules. What's really important in a customer-driven market? We suggest that *delivering with extremely high reliability on promised due dates* is the most important service characteristic of any manufacturing company, and for that reason S-DBR does not schedule the CCR in detail or try to maintain that schedule at all costs. Instead, S-DBR relies on the *master production schedule,* to which even the CCR subordinates itself. Notice that we specified "service" characteristic. We are not ignoring product quality. Rather, we are assuming that there is no shortage of qualified help in assuring product and process quality.*

## Time Buffer

In DBR, a buffer was defined as a liberal estimate of the manufacturing lead time from one control point to another. This is a slightly modified definition—more generalized—than the one given in *Manufacturing at Warp Speed* [Schragenheim and Dettmer, 2001]. In traditional DBR, the control points might be material release and the CCR (for a CCR buffer), or the CCR and the shipping dock (for a shipping buffer). However, in S-DBR the only buffer is the *shipping buffer,* which is defined as a liberal estimate of the time from the release of raw materials until the arrival of the finished order to the shipping dock [Schragenheim and Dettmer, 2001]. The key phrase here is "liberal estimate," which implies that both pure manufacturing time and the appropriate safety time are included. Typically, the size of the shipping buffer is no larger than the quoted lead time (QLT) in make-to-order environments.

If customer delivery-due dates are the most critical service parameter, then for certain they require protection in the form of a shipping buffer. But what about the other two buffers, at the assembly point and at the CCR? Assembly buffers are really no more than additions to shipping buffers, but placed at different locations. The original thinking behind assembly buffers was to ensure that parts of the product that have been worked on by the CCR are not delayed in reaching the shipping dock *after* they leave the CCR. Once we understand that what really matters is preserving delivery on the due dates, then the whole focus is exclusively on timely delivery. Accelerating parts from the CCR, even when there is still ample time until delivery, could result in confused priorities. With an adequately sized overall

---

* Six Sigma experts are readily available for most companies, either internally or from among consultants.

shipping buffer protecting the due date of the complete order, an assembly buffer becomes superfluous.

What about the CCR buffer? Surely a resource as important as the CCR requires specific protection, doesn't it? The answer to this is both "yes" and "no." Yes, it's important to ensure that the CCR is not "starved" for work for extended periods of time. However, when customer demand is always assumed to be the system constraint, a CCR buffer is less critical than the shipping buffer. During those periods when demand exceeds capacity, it becomes more important to minimize times when the CCR has no work to do. After all, as Goldratt observed in *The Goal*, at the constraint, time is money. And time lost at the constraint is money lost to the entire system [Goldratt and Cox, 1986]. However, minimizing the idle time of the CCR doesn't necessarily mean that a predetermined sequence of work at the CCR work must be strictly followed.

Another important ramification from acknowledging the market as the major system constraint is recognizing that some excess, or surge, capacity at the CCR is critical, too. An often overlooked statement by Jonah in *The Goal* is that the flow of work-in-process should be slightly less than demand [Goldratt and Cox, 1986], though there is no hard-and-fast rule. This is another way of saying "Don't fully load your system, even if the demand is there." In fact, Jonah observes that matching the demand on a manufacturing process to the full capacity of its bottleneck is a certain recipe for bankruptcy. The reason is that as a production process begins to approach 100 percent load, backlogs, work-in-process, and eventually late deliveries begin to multiply exponentially. Figure 3.2 illustrates this phenomenon.

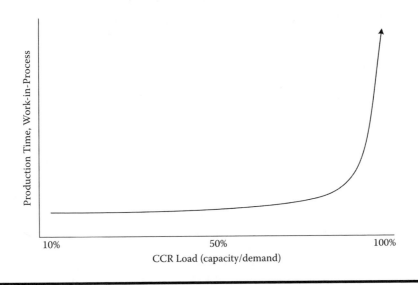

**Figure 3.2    How capacity loading affects production time and work-in-process.**

# Material Release

S-DBR's first task is reducing work-in-process. Doing so facilitates better focus on the really important things and reduces the likelihood of a mistake in production priorities. This is accomplished by forcing all material release to follow a "do-*not*-release-*before*-scheduled-date" requirement. If material release is enforced this way, when we see an order on the floor, we *know* that order is to be delivered sometime between *now* and buffer time from now. No other orders should be on the floor. If delivery is required later than the time buffer from now, it *should not be there*, and you should consider pulling it out of the production sequence temporarily.

Following this practice reduces the WIP in production, preventing the one weakest link from being overwhelmed with unnecessary work. The objective is to maintain only enough WIP in the system that the weakest link can complete within the time buffer. Other work centers should be expected to have *much less* work at their locations because they process work faster than the weakest link.

This is a major precept of constraint theory. To achieve a condition in which most work centers experience limited queues of work, we ingrain the "roadrunner ethic" in the minds of both operators and management. The roadrunner ethic demands that operators work as fast as practical (commensurate with performing quality work) when there is work to do. However, they must not create work when it isn't necessary. In other words, work the way firemen do: When there's a fire, be fast and efficient. When no fires are reported, *don't create one*—please! However, we can expect operators to behave that way *only* when they are absolutely assured of not being the victims of "efficiency."

When the *rope* is connected to actual market demand, management's challenge is to find a way to keep the CCR from being either overloaded or starved under dynamic conditions that may change daily. Applying a relatively constant time buffer becomes less effective, but trying to make the buffer as dynamic as the changes in demand is not easy. S-DBR suggests a slightly different approach—planned load.

## *Planned Load*

Planned load is a critical concept in applying S-DBR and in actually being able to quote reliable delivery dates. The planned load for a resource was originally defined as the total number of hours required to complete all work that has been formally released into the system [Schragenheim and Dettmer, 2001]. Clearly the resource of primary interest is the slowest one—the CCR—or those resources that may become the CCR. Instead of maintaining—and trying to continually adjust—a CCR buffer, why not consider the total demand for the CCR's time during the period approximating the lead time promised to the customer? In other words, monitor the total load on the CCR as it changes daily to ensure that it never exceeds some arbitrary value (less than 100 percent, but still reasonably high—perhaps 80 to 90 percent). S-DBR calls for metering demand (the acceptance of customer orders); this is done on top of choking the material release based on customer due date minus shipping buffer.

How does the demand control happen? Refer to Figure 3.3. Within the standard quoted lead time, the CCR is capable of a finite number of hours of work. Every new work order that comes into the system consumes a block of that capacity, until some limit is reached. To prevent overloading the CCR, we set that limit at somewhat *less* than 100 percent of capacity. The actual limit will vary from one situation to another. Until that limit is reached, there is more capacity than demand, as reflected in the middle of the demand curve in Figure 3.3. However, when demand exceeds capacity (the left and right sides of the curve), we must restrict the admission of new orders to prevent overloading the CCR. (A variety of tactics is available to achieve this; for example, quoting a longer lead time when the order is received or raising prices to depress demand slightly.)

The net effect, however, is to avoid the need for a CCR buffer. The CCR will normally have enough work waiting for its attention most of the time because it is, after all, the slowest work center. At the same time, the one and only buffer—at the shipping dock—ensures that delivery to the customer is made on time.

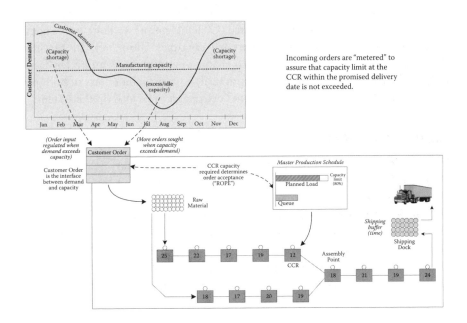

**Figure 3.3   Planned load: Controlling the CCR's workload.**

## Significance of Planned Load

Consider this analogy. Who among us has never been stuck in a traffic jam on a freeway or other limited-access highway? When such traffic jams occur, flow often grinds to a halt. Clearly there are too many cars. But how much of a reduction in cars on such a highway is required to get traffic flowing again? Fifty percent? Hardly. Often a reduction of no more than 10 percent in the number of cars is enough to restore traffic flow to near the speed limit.*

What threatens success in meeting a delivery due date is lack of capacity at a specific resource, the weakest link in the production chain, the resource that has the least excess capacity. In other words, this resource would probably become the system constraint if demand grew substantially. Moreover, it would not even have to reach 100 percent of capacity consumed for that to happen. Remember the example of the congested highway described above: a resource might be fully loaded (or overloaded) and somewhat less than 100 percent of its apparent capacity. This is why the weakest link's planned load must be carefully monitored to determine whether capacity shortage—one that would compromise delivery times—is imminent.

Executing the planned-load concept effectively requires keeping a close watch on the scheduled work compared with the available capacity for the period of the lead time promised to customers. The planned load on the weakest link is the most important parameter to keep an eye on. When that load approaches the standard lead-time our customers expect from us, then the weakest link is on the verge of becoming a capacity-constrained resource (CCR). It means we don't have enough excess capacity at the CCR to quote short delivery times. Ultimately, this precludes obtaining more business or being able to charge premium prices for better (faster) service. The role of the planned load is primarily to assist joint efforts of both production *and* sales to assure the maximum "deliverable" real load on the system.

## Planned Load Benefits

If we monitor the planned load effectively, we realize two primary benefits. The first is very early warning that new orders to be received would likely be late if the regular lead time is applied. If we know at the time the order is received that the planned load on the CCR is well within its capacity (say, 80 percent or less), then the CCR will never cause a bottleneck under normal circumstances. If the planned load on the CCR rises above a safe limit, this is known *at the time the order is received*, and there is no need to wait for the order to be released, and certainly no need to wait until it reaches a red zone to know that it's going to be late if we don't do something immediately. This signal is provided at the time the order enters the logbook.

---

* Just ask any resident of Los Angeles, California, who has seen traffic flow increase after a blockage, often to 75 miles per hour or more—well above the posted speed limit—while spacing between cars remains frighteningly small.

In traditional DBR, the early warning signal would also occur, but not until the subsequent schedule run that would include the specific order and others that have been accepted since the last schedule. Thus, in traditional DBR, it's often necessary to recompute the schedule frequently just to know whether newly received orders have a good chance of on-time delivery or some risk of being late.

The second benefit of carefully monitoring planned load is simplifying the alert system to a single warning. If a process should break down at any point, the small amount of excess capacity at the CCR (and considerably more capacity everywhere else in the system) will normally be enough to restore the delayed work to "on schedule" without extraordinary intervention by managers or supervisors. In traditional DBR, the schedule of the CCR would have to be replanned to allow other orders to use the time planned for the broken process, and to reintroduce the delayed order later on. Having to monitor three buffers also adds to the amount of changes resulting from one order that must be delayed. The three-buffer system doesn't provide the flexibility required to deal with all of the orders, because consideration of the impact of a red order in the CCR buffer to a red order on an assembly buffer isn't straightforward. The three different kinds of buffer penetrations are much more complicated to manage.

## Simpler Buffer Management

Paying special attention to the planned load for the CCR (to keep it well below 100 percent of capacity) sets the stage for uncomplicated buffer management. When an order can be found in three different priority lists, the result is confusing priorities. For instance, an order that is in the red zone for the CCR buffer might be in the green zone for the shipping buffer. This is not uncommon, and we could certainly explain why it happens, but having to do so adds unnecessary confusion. Moreover, suppose a work center discovers three orders in red, but one of these orders is red with respect to the CCR buffer. Another order is in the red for the assembly buffer, and the third is red for the shipping buffer. Which of the three requires immediate attention? This confusion doesn't even exist in S-DBR because there is only one buffer to monitor, and it reveals all the relevant information regarding its relative priority.

In the Management Interactive Case Study Simulator (MICSS), we introduced a simplified buffer management—orders that penetrate the "red line time" are red orders, all the rest are just "not red" and don't warrant special priority. When the red zone is penetrated, the simpler single warning means that an order's due date is at risk. Ideally, we would recognize this signal in sufficient time to take corrective action and avoid a late delivery. Buffer management allows for three categories of priorities by adding the yellow zone, recognizing that a yellow zone order assuredly has a higher priority than a green order, but not as high as a red zone order.

When there are too many red orders, the conclusion should be that the system's stability is threatened—typically caused by the emergence of a new CCR. This method of red line control can also identify a new CCR by highlighting the resource with the majority of red orders backlogged behind it.

# S-DBR: Beyond the Basics

Since the introduction of S-DBR, many lessons have been learned as it has been implemented in many types of plants. In particular, we will examine what we've observed about S-DBR relative to earlier and contemporary schemes for shop floor control. Understanding the nuances associated with a new approach requires us to examine how S-DBR compares to the other systems as well as identify features that both clarify and simplify the task of managing when S-DBR is employed.

## S-DBR and MRP

At its most basic, the rationale behind Manufacturing Resource Planning (MRP) is somewhat similar to S-DBR. MRP starts with the bill-of-materials of an order. Each part in the bill-of-materials has its own lead time. From the due date, MRP calculates the starting time for the final assembly/testing/packaging, and then proceeds down a bill-of-material "tree" to component parts required for the higher-level part. At the "root of the tree," need dates for all materials are determined. Different materials required for a complicated order might be assigned different need dates, depending on the number of levels in the bill-of-materials that the materials are actually required to pass through. This aspect of MRP differs from S-DBR, in which all the materials required for an order are assigned the same release date. But it's similar to S-DBR in that capacity requirements are not fully considered at the time of the basic planning. S-DBR, too, is concerned with the CCR's capacity requirements, but the capacity of most resources is ignored. Thus, in S-DBR the same product always gets the same production lead time, which includes the time from material release until finished product completion.

Another critical difference between MRP and S-DBR is that in MRP every customer order is divided into several (sometimes many) work orders, each associated with one part in the bill-of-material (BOM) and the processes required to produce it from the lower level parts in the BOM. Every individual work order, representing just a fraction of what is required for the complete order, merits an intermediate due date. In contrast, S-DBR considers everything that is required to complete a customer order. No intermediate due dates are assigned—just one due date for the order completion. The presence of intermediate due dates could prompt an attitude that "there is no need to rush this order, the intermediate due date is not too close." In other words, the time buffer is usually wasted at operations that already have enough excess capacity to just keep the order moving, but when something goes wrong, too little time remains to recover. The S-DBR philosophy is that the time assigned includes *all* operations, so every order must be moved as quickly as possible. And buffer management helps pinpoint the right priorities when several orders arrive simultaneously at a work center.

Even though the basic idea behind MRP is simple and straightforward, eventually the MRP algorithm becomes complicated and confused. Because the same part might be required for different customer orders, MRP combines several work orders for a particular part together into one work order for the common part. This causes loss of direct visibility for which customer sales might be impacted by a delay in processing that order, and which ones would not impacted. Batching policies make MRP even more complicated.

Another source of confusion is that MRP treats the material release schedule the same as any other schedule. That is, it represents a not-later-than date. Therefore, if the resource in the first production operation has nothing to do at the moment, most production managers release the materials earlier than MRP dictates. As a result, the shop floor is filled with orders equivalent to the full capacity of its gating operations (the first operations to process orders). Batching policies, combined with the reliance on forecasting to produce based on expectation rather than solid demand, cause make-to-order work orders to be combined with make-to-stock orders. Mixing production of parts for defined customer commitments with production of "maybe-we'll-have-a-demand-soon" orders adds another level of complication and confusion. These complications basically turned a *pull* methodology into *push*.

The fundamental deficiency inherent in MRP is that it fails to consider capacity to be finite. It's not uncommon to find certain resources overloaded at the exact time when an order for Product X should be processed. In many situations this kind of overload condition can delay customer delivery. Ironically, because MRP gives intermediate due dates to every work order, the temporary overload for a resource causes a delayed delivery, yet S-DBR experiencing the same overload would still ship on time.

Here is an example to illustrate this situation. Suppose that in MRP one particular customer order contains three work orders: Work Order (WO) 1 requires part A to be finished in a week. Then WO 2 takes part A and makes part B—also a week. Then WO 3 takes part B and completes the customer order in yet another week. The production lead time, at least on paper, is three weeks.

Suppose now that part A is completed on time—actually it's finished two days ahead of time, but it waits to be delivered to the next step. Now, a specific resource required by WO 2 has an overload of other work that delays its availability by two weeks. That specific order is delayed for a whole week. Consequently, when Part B becomes available for WO 3, the due date is already here. The people responsible for WO 3 decided to do other orders that could be delivered on time, rather than expediting this order. So at the end, the order was delivered late by one week.

What would have happened if S-DBR had governed the production process? First, only one order would have been defined—the customer order. The shipping buffer could be three weeks, or even less. The order is expected to flow as fast as possible through all work centers. But one work center has an overload, so some orders could be late. But which ones? S-DBR prescribes referring to the buffer management list and letting orders in the red zone take priority. Suppose at that time one-third of the buffer of the order in question has been consumed—it's in the yellow zone, meaning it should have priority over green-zone orders. Now suppose this didn't help; there's a delay of a week, and that order is now in the red zone, but it's downstrean of the work center causing the blockage. The order is expedited by all subsequent work centers, which means it receives the highest priority everywhere. Since all of these downstream work centers have sufficient protective capacity and processing time is relatively short, is there any reason why the due date should be missed?

The point is that in S-DBR the time buffer protects the *whole production process*. Thus, if one operation is delayed, possibly due to lack of capacity right at the moment, the overall time buffer is usually enough to offset the delay. When a late start at a work center is detected, it's possible to take some remedial actions immediately—adding overtime, for example. Moreover, because an order's penetration of a red buffer zone also triggers a higher priority for that order, it would be worked earlier than other orders. By reducing the waiting time for that order, there is still a reasonable chance that the order will be completed on time.

## S-DBR and Lean

As we saw in Chapter 2, the Japanese forced a rapid paradigm shift in manufacturing concepts with the introduction of the Toyota Production System (TPS), an approach that morphed and was adopted in the West as "lean." One of the key components of the TPS/lean concept was just-in-time (JIT), a set of production rules designed to make production processes faster, more flexible, and less clogged with work-in-process [Womack and Jones, 1996].

Drum–buffer–rope, simultaneously conceived by Goldratt before just-in-time really caught on in the West, aimed to do much the same thing, but with one key difference. DBR recognized that management consultant and author Joseph Juran was essentially right: there is a practical (and economic) limit to how much process and product variability can be eliminated from any process (Juran and Gryna, 1988). In other words, at some point the cost of searching out and correcting quality defects is exceeded by the costs of defect prevention. For most production processes, this still leaves some degree of general-cause variation in the system, and there still remains the issue of special-cause variation ("acts of God").

So while the near perfection of Six Sigma quality is an admirable goal, the reality is that most companies are unlikely ever to reach it. Thus, some tolerance for variability in manufacturing is both desirable and necessary. JIT/lean made informal allowances or adjustments for variation that could not be completely eliminated. DBR formalized those allowances in the form of buffer management. Both methods sought to manage flow of work through the manufacturing process. DBR recognized that such flow would be limited by the least-capable resource and accepted some excess capacity in most places. In the interest of increased local efficiency, JIT/lean sought to balance capacity to the degree possible so as to eliminate waste.

Even when the internal variations are minimized, the variations in the market demand can be very large. Customer demand directly impacts the load on the key resources, notably the CCR, which immediately impacts the actual production lead time. The time buffers used in S-DBR are capable of protecting the due date performance against variations in demand, which cannot be accommodated by Six Sigma. The combination of variation in the demand and internal variation, even when the latter is relatively small, makes S-DBR a preferred approach.

Some additional differences between Lean/TPS/JIT and S-DBR are in the general approach to change and in the way WIP is spread through the shop floor. Theory of Constraints (TOC) in general strives for the minimum change required to achieve breakthrough results. This allows S-DBR implementations to be fairly fast. The S-DBR approach is certainly very "lean" by advocating perfect "pull." When the order has been received, given a due date, and the due date minus the time buffer is today, only then are the materials for that order released to the floor. The manufacturing process has not changed. The only notable difference observable in the execution phase is buffer management; orders are colored based on how urgent they are.

The other change is, of course, the spread of inventory through the shop. The time protection is targeted on the due dates. The relative shortage of excess capacity at the CCR causes more WIP to accumulate there. In TOC, it makes sense to have one focus for management attention. Ultimately, not all the work centers are equal, and it is not in the interest of the company to make them artificially equal.

## *Reducing a Buffer*

What happens when the time buffer is very long relative to its "optimum" size? The longer the buffer, assuming it applies to many orders, then the higher WIP on the floor will be (more orders are released). The higher the WIP, the longer the actual production lead time because the wait time is longer.

Our recommendation is to take a bold step: determine a time buffer that is significantly shorter than the current lead time (before S-DBR has been applied). Significantly shorter means 50 percent of the current lead time.

Once the new time buffer has been determined, it should automatically be applied to a specific order. Materials for an order of Product X are then released for production at the shipping buffer time before the due date. Once the order has been released, buffer management, the corresponding control mechanism prescribed in TOC, shows us the relative priority of the order based on actual time consumed during production compared with the predetermined shipping (time) buffer. Therefore, this order is now competing with the other orders for the available capacity of all the required resources. The priorities suggested by buffer management determine which order gets the available capacity now. Of course, as time progresses and the priority of the specific order increases, that order would spend very little time waiting for any resource because its priority is now the highest.

## Implications of Dealing with Market Constraints

A funny thing happened on the way to greater efficiency. While major efforts to manage the constraint might be extremely successful, the "conquered" constraint may in fact be able to produce whatever the market demands. Customer (market) demand becomes the major constraint for most manufacturing companies, not the ability to produce. Actually, this is an inaccurate way to express the situation. The market constraint has always been present. It is merely masked in the short term by an apparent constraint in production. Goldratt expressed this condition very well in *The Goal* [Goldratt and Cox, 1986]. After eliminating the production bottleneck in his plant, the chief character finds that he doesn't have enough demand for products to fill up his capacity, so he presses the vice president of sales to generate more demand.

By the mid-1990s, a critical mass of companies had embraced lean and, excluding those that botched its implementation, had gotten what they wished for: less WIP and shorter production cycles. However, they also got something they hadn't planned on: idle capacity because the improved use of their resources "liberated" hidden capacity.

For many companies, the solution to this new problem was obvious, and somewhat short sighted: trim excess capacity. In other words, get rid of wasted (idle) capacity. Combined with a surge in moving manufacturing offshore, the increased efficiencies delivered by lean made excess capacity seem unnecessary. Rather than bending efforts to exploit this new constraint with accelerated sales efforts, most companies tended to lay off employees and shut down or sell off facilities. Practitioners of DBR, while smarter than their lean-practicing contemporaries concerning the need for excess capacity in their plants, still experienced the same concerns—not enough demand to fill up available capacity.

However, the market constraint is not really a new constraint. It has always been there. Until we improved our production process, through either DBR or lean, it just wasn't visible. Once it becomes visible, a market constraint demands attention. The common solution, as suggested earlier, was to "eliminate half the glass." But this is not a particularly good prescription for corporate growth. It's more appropriate for a company that is fighting a defensive, or reactive, battle. And who wants to be *that* company?

Even during times of fully loaded or overloaded capacity, when it seems that a company's constraint is definitely internal, as long as untapped customers still exist out there, the market can be considered a constraint. It just isn't immediately visible because of the urgent pressure of the temporary capacity overload. Now that the market is correctly identified as our long-term constraint, we ought to develop a plan to exploit it and then subordinate our production process to the market.

1. **Identify** the system constraint
2. Decide how to **Exploit** the system constraint
3. **Subordinate** everything else to the decision in Step 2
4. **Elevate** the system constraint
5. Go back to Step 1, but avoid **Inertia**

**Figure 3.4   The five focusing steps.**

## S-DBR and Market Exploitation

The next logical question is: "How does S-DBR support exploitation of the market"? The answer is through subordination—Step 3 in the Five Focusing Steps presented in Figure 3.4. Instead of making internal efficiency the ultimate objective of the production process, we make short lead times and consistent on-time delivery reliability the objective. Naturally, this means that we must be less concerned with efficiency at any given resource in the production process, or the production process as a whole. We don't disregard efficiency; we just don't expend extraordinary effort to improve it. In this respect, DBR (both simplified and traditional) diverge substantially from lean manufacturing.

What, then, does an S-DBR production line, appropriately subordinated to the market, look like? Figure 3.5 illustrates the S-DBR concept. New orders are received and daily assigned due dates based on the existing shipping buffer length. A feasibility check is then accomplished using the planned load. The new orders are added into the planned load, while orders that have been completed that day by the CCR are deleted from the planned load. The planned load time is then compared with the standard quoted time. When the updated planned load seems reasonably safe, meaning that there is enough time between the value of the planned load and the standard quoted lead time, the new orders become customer orders. The material release date is then determined, based on the due date. If the comparison shows an under-capacity, two valid alternatives can be considered: either obtain additional capacity (overtime, outsourcing) and confirm the due dates, or advise the customers immediately of extended promised delivery dates. If the delivery dates are pushed back, a change in the standard lead time is required, at least for the short term, reflecting the peak in demand. Execution is subordinated by the simple, straightforward use of buffer management.

Focusing exclusively on subordination to the market (meeting customer commitments) clearly differentiates S-DBR from earlier practice of DBR, in which the objective of subordination was not to miss out on even one minute of CCR capacity. The second of the Five Focusing Steps is verbalized: "*Decide* how to exploit …", not simply "exploit." Goldratt has often said, "Don't be too greedy in exploiting the CCR—leave something on the table." Nevertheless, most TOC practitioners have historically tried to squeeze every bit of capacity out of the CCR.

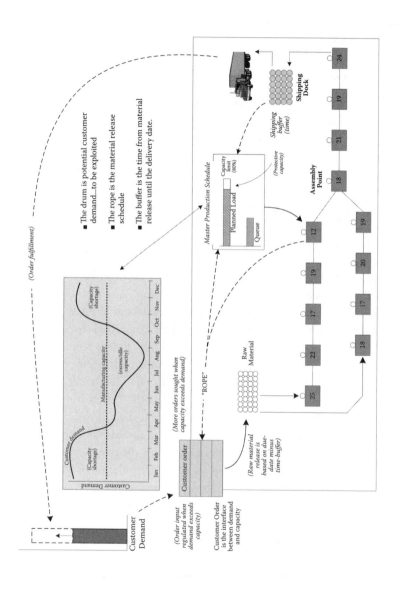

**Figure 3.5  A production process effectively subordinated to the market (S-DBR).**

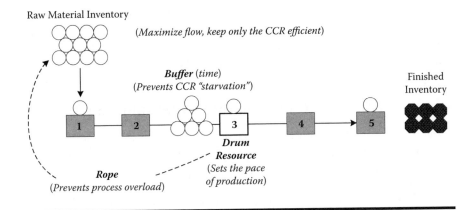

**Figure 3.6   Traditional drum–buffer–rope manufacturing process.**

Now, if you do that (extract every last bit of an internal resource's capacity), the inevitable result is that some customers pay the price, either through excessively long lead times or by missed original due dates. This could easily happen because the CCR buffer doesn't protect against Murphy's law affecting the CCR itself. Consequently, if the CCR schedule is exceptionally tight for, say, the next three weeks (every bit of capacity is squeezed out of it), what is scheduled to be produced three weeks from now has a high probability of being delayed to some degree. This means that the shipping buffer—the time given to the parts to move from the CCR to the end of the process—also has to absorb any delays caused by the CCR itself. The longer the horizon of the CCR schedule, the more problematic this issue will become. Late deliveries induced by the CCR will be more common.

Of course, delays in CCR processing would impact the planned load in much the same way that they impact the CCR finite-capacity schedule. But there are key differences. First, the planned load does not impose a sequence on orders; priorities are determined by the market requirements. Second, the planned load is monitored so it will always leave enough time to compensate for such delays. Finally, the planning time horizon is limited to the standard lead time because in S-DBR the objective is not to squeeze every bit of capacity, but rather to meet all the commitments made to the market.

Clearly, this is a radical departure from traditional DBR, which is reflected in Figure 3.6. Traditional DBR was designed exclusively to manage the smooth flow of work through the production process, from material release to completion of finished product. As characterized in *The Goal* [Goldratt and Cox, 1986] and described in *The Race* [Goldratt and Fox, 1987], smooth, rapid flow through the system was strictly a function of flow through the bottleneck, or CCR. DBR maximized this flow of work through the CCR without creating a backlog. The optimum pace of the CCR determined the pace of the entire production process. Similar to the boy-scout hike described in *The Goal* [Goldratt and Cox, 1986], the rest of the production process (which is presumably more capable, or faster, than the CCR) is "throttled back" to match the pace of the CCR, which determines the rate at which materials are released into production. This constitutes the "rope," which is "tied" between the CCR and the material release point.

## Managing the 'CCR' in a Market Constraint Situation

If our new worldview considers market demand to be the constraint, and our exploitation plan is to offer the shortest, most reliable promised delivery dates, what happens to our traditional concept of a CCR? We know from our experience with traditional DBR that a CCR is the resource in a production chain with the least capacity. In a market constraint condition, that's still a valid definition for S-DBR. It's just less crucial because most of the time demand is lower than the full capacity of the CCR, so it is seldom backlogged with WIP. It's clear that a CCR exists whether it's either *active* or *inactive*. It's active during those peak times when demand is close to, or even exceeds, the CCR's capacity; it's inactive most of the rest of the time. Yet it still potentially restricts our ability to exploit the market constraint to the fullest extent, so we must be concerned about it.

In a traditional DBR situation, we scheduled and protected the CCR to squeeze as much throughput[*] out of it as possible. We treated all other resources as non-constraints (non-CCRs). This means that we accepted the fact of excess capacity at these resources. Even more, we *sought to ensure* that they always had excess capacity, so that we would not experience interactive constraints and the chaos this might cause. We monitored the CCR constantly to make certain it would not be overloaded, and we watched non-CCRs less frequently to ensure that changes in market demand or internal problems didn't make a different resource into a CCR.

Subordinating to a market constraint essentially requires raising the same thinking to a higher level. When we have a considerable amount of idle capacity everywhere in our production process, all internal resources are nonconstraints. One of these resources in the chain is still the weakest link—it just is not an active CCR at the moment. However, it could be, and will be, if demand should suddenly peak for any reason. What reason could there be for such a peak? How about promising earlier and more reliable delivery dates than any of our competitors? Such an exploitation strategy can win business from competitors. Isn't it then likely that this new business could raise demand above the level of our internal CCR?

Thus, in a market constraint condition, we must continually exploit demand to keep sales up, yet at the same time we must be careful not to activate an internal constraint that could compromise our ability to meet delivery promises to our

---

[*] Throughput is defined as the rate at which the organization generates units of whatever goal it strives to achieve. In business organizations it is considered revenues minus the truly variable costs. Refer to Chapter 13 of *Manufacturing at Warp Speed* (Schragenheim and Dettmer, 2001) to understand the term and its impact.

customers.* Besides delivering a shoddy product, the next-fastest way to alienate customers is to inconvenience them by failing to deliver when promised. S-DBR is well suited to exploiting the market constraint.

## Summary

Traditional DBR came onto the scene at about the same time that just-in-time (a component of the Toyota Production System) became widespread in the West. The two methodologies shared similar characteristics (e.g., minimizing inventory in production, increasing flow rate, decreasing manufacturing cycle times). But whereas JIT, and later lean manufacturing, focus on waste elimination and maximizing efficiency everywhere in production, DBR's objective has always been shorter, more reliable delivery times. Some of what lean adherents would characterize as "waste" is tolerated in DBR as a way of ensuring high on-time delivery performance under conditions of uncertainty that are often challenging for lean manufacturers to accomplish.

S-DBR represents the next generation of DBR. It treats the production process as part of a larger system—the business enterprise. It integrates demand generation (sales and marketing) with production planning and management by pacing acceptance of customer orders with CCR capacity. It accommodates uncertainty and variation as effectively as traditional DBR does, but it requires monitoring far fewer performance metrics to ensure success. In this respect, it is inherently simpler than traditional DBR and certainly more so than MRP-II production.

S-DBR's production policies may be summarized as follows. (For a more detailed explanation, refer to *Manufacturing at Warp Speed*, Chapter 10 and Chapter 11 [Schragenheim and Dettmer, 2001].)

■ Production planning is accomplished daily, incorporating the previous day's information, updated by the deletion of work completed and the addition of new orders received since the last plan.

■ Raw material is not released to the production floor before the due date minus time buffer for the order.

■ Order acceptance is always conditioned on planned load for the CCR.

■ Minimum process-batch size is tolerated only when a resource becoming a bottleneck is a real possibility.

■ Red orders are top priority. Conscientious buffer management also requires a higher priority for yellow orders. In all other situations, operators decide priorities among same-color orders.

---

* If you want a glaring example of what happens to a company that fails to deliver expected service to its customers, just look at most commercial airlines in the United States. With very few exceptions, opinions are unfavorable.

- Transfer of work-in-process between workstations will be made in the small-est-sized batch practical (one piece, if possible).
- The shipping buffer includes safety time sufficient to remediate the most likely reasons for production stoppages anywhere in the production process. A "red zone" will be established equal to one-third of the shipping buffer.

In the next chapter, we will examine some of the advances and refinements in S-DBR, with an emphasis on using it in make-to-stock conditions.

## References

Goldratt, E.M., and Jeff Cox, *The Goal: A Process of Ongoing Improvement.* Great Barrington, MA: The North River Press, 1986.

Goldratt, E.M., and Robert E. Fox, *The Race.* Great Barrington, MA: The North River Press, 1987.

Juran, J.M., and F.M. Guyna, *Juran's Quality Control Handbook* (4th ed.) New York: McGraw-Hill, 1988.

Schragenheim, E., and H.W. Dettmer, *Manufacturing at Warp Speed.* Boca Raton, FL: St. Lucie Press, 2001.

Womack, James P., and Daniel T. Jones, *Lean Thinking: Banish Waste and Create Wealth in Your Corporation.* New York: Simon & Schuster, 1996.

# Chapter 4

## Enhancements to Simplified Drum–Buffer–Rope

## Contents

Since the publication of *Manufacturing at Warp Speed*, S-DBR has matured significantly. In the latter part of the last chapter, we pointed out that some of the developments since the introduction of S-DBR have added significantly to its robustness. What follows is a deeper discussion of some important supporting elements of S-DBR, such as how to estimate safe delivery dates and how safe dates affect other measures. The better we schedule, organize, and synchronize based on realistic safe dates, the more we can realize benefits from a planned load. We will also present a more comprehensive explanation of issues related to the effective management of capacity. Finally, we will examine some of the issues related to reliability and value that are somewhat unique to S-DBR.

## Safe Dates: Interfacing Sales and Production

S-DBR, as alluded to in the previous chapter, had only an indirect way of checking for possible overload of the capacity-constrained resource (CCR). This procedure was validation that the planned load is not too close to the standard lead time. In other words, the planned load provides a warning signal that we are about to run out of capacity. This is a different signal than the red-line warning we would observe if the shipping buffer was practically exhausted. The planned load warning indicates that we face considerable risk of a late delivery of one or more orders. Buffer management, the generic red-line control methodology, should also be used to provide warning signals whenever the *rate* of red-zone orders exceeds 10 percent of the *total number* of orders. However, this warning might come too late to take effective corrective action. So the planned-load warning should be carefully monitored to prevent any embarrassing situations.

This situation assumes that the sales department continues to generate new orders and quote delivery dates without consulting operations very much (if at all). The only time that actions to hold back sales are applied is likely when production starts experiencing a fast onset of overload. At that time, sales might quote longer delivery times or even increase the quoted price, but only after a lot of complaining by production.

An alternative is for production, each day, to provide sales with highly reliable "safe dates" for promising delivery of the orders that sales might close during the next day. Sales could—and actually should—quote even later dates, but production's commitment to deliver every order by a safe date (or earlier) would be an asset for the company as a whole. Certainly, "safe dates" should be determined with the objective of properly utilizing the available capacity. But think how advantageous it could be to achieve near-perfect delivery reliability in the shortest practical time. Figure 4.1 shows the current state of planned load for a CCR.

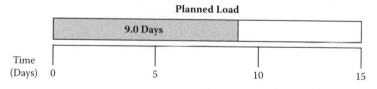

Figure 4.1   Planned load: An example.

Based on the number of units the CCR is required to process and the available working hours per day, all the orders to be shipped within the horizon of some designated lead time are considered in the calculation of the planned load. This designated lead time may be typical of other competitors in the industry, or it could be some arbitrary figure. The planned load considers only the CCR time required to process these units. Let's assume that a standard lead time is 15 days. If we are now at the start of "day 1," the CCR will have to work continuously for nine full days to complete all of the confirmed orders already introduced into production.

If a new order were to be received now, on what day do you think the production manager could feel reasonably confident in promising that it would be shipped? The planned load provides the most relevant information. We may assume that the new order will start processing at the CCR sometime around day 9. It might be earlier, if the materials arrive there and the operator decides to prioritize that order before other orders introduced ahead of it.* It also might be later than the ninth day if the CCR experiences idle time during the first eight days, or if the materials for the new order happen to arrive late. However, all other things being equal, we would expect processing of the new order to begin on or about day 9. If a serious delay were to occur, buffer management would provide an "expedite" signal, and the order would be rushed to completion.

This is good, relevant information—not very precise, but sufficient for our needs. But this is not all the information we need. We must also know how long it might take for an order to pass through the CCR and all the subsequent operations before arriving at the shipping dock. And that time must *not* be an average. It has to be longer than that to allow for various kinds of delays that might occur *after* the CCR, so as not to interfere with timely delivery. In other words, we need to keep in mind that part of the buffer is required to protect the part of production from the CCR to the end. This means we must know:

- The shipping buffer (meaning the time it could take from material release until completion)
- The location of the CCR in the operation

Let's make a "typical Theory of Constraints (TOC) decision." We will assume that adding half of the shipping buffer to the planned load is a close enough estimate of the time required for an order to pass through the CCR and all the way to completion. If the CCR is roughly halfway through the production process, this assumption also means that it will take up to half of the shipping buffer to move an order from material release until it reaches the CCR, ready for processing. In the example shown in Figure 4.1, suppose the shipping buffer is estimated to be 10

---

* As long as the Red–Yellow–Green concept is applied, the operator has the latitude to change normal priorities; for instance, in order to save setup time by combining like orders, or because materials for other orders have been slightly delayed.

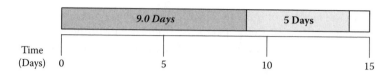

**Figure 4.2  Determining a "safe date."**

days. If we add half of that, our safe date should be the 15ᵗʰ day (the end of day 14 or at the morning of day 15) as shown in Figure 4.2.

This "safe date" can be considered a conservatively safe commitment. It's based on the approximate determination of the capacity required for earlier orders, plus a liberal estimate of the time required for an order to pass through the CCR operations and all the remaining steps to completion. Notice that, especially after the CCR operations, the wait time is minimal because the relatively slower pace of the CCR does cause a great deal of WIP to accumulate in the former part of the process, while after the CCR the pace of the flow is under the full dominance of the CCR, and the faster resources simply flow the orders out.

## *Estimating Release Time Based on a Safe Date*

Once we have a safe date, when should we schedule the release of materials? S-DBR guidelines suggest *safe date minus the shipping buffer time*. That means we release the materials half of the shipping buffer time prior to the date of the outer boundary of the planned load. For example, if we have a full shipping buffer of 10 days as shown in Figure 4.3a and we are deciding the release date for materials, we may subtract the shipping buffer from the safe date. Alternatively, we can subtract half of the full buffer (five days) from the planned load. Either method determines that we should release materials on day 4 as shown in Figure 4.3b.

From the procedure described above, we obtain the following information:

■ A means to convey (and convince sales) the earliest date that production can guarantee delivery.

■ An accurate picture of the opportunities and the threats. The time of the safe date depends on the current load in production. In other words, in some situations we could offer a very early safe date, which could initiate certain opportunities. On the other hand, sometimes it would be a somewhat later date because of the orders with committed due dates already assigned.

■ We now have a way to *smooth the load on the shop floor*. By providing safe dates, we ensure no temporary large peak loads. When a "spike" in orders occurs, subsequent orders are deferred a little later in time.

■ This procedure essentially ties the rope to the planned load. Thus, the amount of WIP in the production remains relatively low. The WIP still keeps the

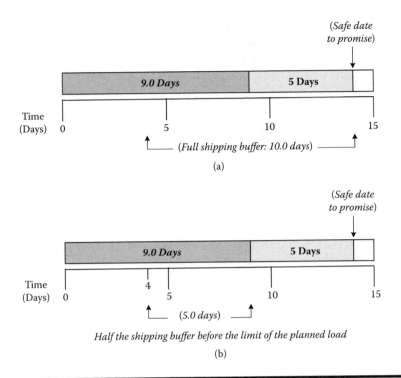

**Figure 4.3 When to release materials.**

CCR as busy as possible without overloading it. Wait time is minimized without wasting the CCR's valuable capacity. All this permits a very short shipping buffer time.

The last point is especially important. First, it clarifies the distinction between the production lead time and the quoted lead time. Because of the change in the perception of production lead time being just part of the lead time offered to potential clients, it is suggested that we change the name "shipping buffer" (the liberal estimation of the time from material release and completion) to "production buffer." The change in name highlights the fact that production buffer is the time for production alone, but it could actually be longer (in calendar time) because of the time until the order is actually released to the floor.

## Quoted Lead Time and Safe Date

The quoted lead time is *the time the client must wait after placing the order until receiving it*. In make-to-order situations, production lead time (what we previously called the shipping buffer) is included in that time. So the question becomes, how long is the production lead time compared with the quoted lead time? The procedure described above indicates that it could be a fairly small part of the overall quoted lead time. An order just received might wait a while before it's released to the production floor. This wait time "off the shop floor" depends entirely on the current planned load. Holding the order before releasing it helps maintain production's attention on the execution of truly important orders, and it facilitates the quick flow of orders through production.

While this procedure establishes safe dates, it doesn't necessarily mean the safe date will be the actual due dates. The reason is based on sound marketing principles. If providing added value (i.e., shorter deliver dates) for free is not likely to be beneficial, then sales should quote the longer standard lead time every time the safe date is prior or equal to that date.

When the standard time is significantly longer than the safe date (and we have no compelling reason to offer the earlier date), when should we release the materials? Let's look at a specific example. Table 4.1 includes a list of XYZ's current orders and relevant production information.

Assume that at XYZ Inc.:

■ Standard lead time in the industry is 30 days.
■ There is one CCR working 24 hours a day, 7 days a week.
■ At the moment, our company has already committed to the following orders (CCR time includes setup time). (See Table 4.1)
■ A new order arrives for 70 units of P3. Estimated CCR time (including setup) is 24 hours.
■ The production buffer is established at 10 days (240 hours).
■ The total existing planned load (excluding the new incoming order) amounts to 186 hours (7 days and 18 hours from now).

**Table 4.1   Committed Orders**

| Order | Product | No. of Units | CCR time req. |
|-------|---------|--------------|---------------|
| 1 | P1 | 40 | 12 hours |
| 2 | P6 | 100 | 18 hours |
| 3 | P7 | 90 | 12 hours |
| 4 | P15 | 40 | 24 hours |
| 5 | P1 | 100 | 24 hours |
| 6 | P2 | 40 | 12 hours |
| 7 | P6 | 50 | 12 hours |
| 8 | P1 | 100 | 24 hours |
| 9 | P12 | 40 | 36 hours |
| 10 | P10 | 100 | 12 hours |

- The proposed due date is: 186 + 240/2 = 306 hours from now, or 13 days from now (rounded up to the nearest whole day).

These are the facts. But should we propose delivery in 13 days just because we can do it this time? If a similar order arrives two months later, and if demand has more than tripled during that time (planned load increases to 25 days), then quoting the regular 30 days will be precisely the situation in which the safe date is the same as the standard time.

Salespeople are always eager to nail down a sale. When the industry standard lead time is 30 days and demand is relatively low, marketing perceives the opportunity to offer delivery in 13 days as exceptionally good. But what kind of expectations would this instill in the client in the future? Would that client expect the same fast response every time? It is not certain that the company can make good on such a promise all the time.

Moreover, for some clients a delivery in less than half the standard industry lead-time could be very beneficial, so much so that they might be prepared to pay considerably more for a guaranteed delivery in that time. However, if from time to time you routinely agree to deliver in half the standard time for the regular price, how can you ever expect to ask (and receive) a higher price for doing so? Our point is that the ability to guarantee delivery in substantially shorter time than the competition is a competitive advantage that can translate to higher-than-normal revenues, *if* it is offered judiciously. You become the preferred supplier when the stakes are high for the client.

Shortly we will discuss the opportunity to offer faster response times to the market for a significant markup. But for now it's sufficient to recognize the ramifications of quoting the standard time—in our example, 30 days from now. The practical question is: When should we release the order, received now, when the planned load is 186 hours and the production buffer is 10 days (240 hours)? Refer to Figure 4.4.

The typical S-DBR procedure suggests releasing the order at 186 hours minus 120 hours, or about three days from now. But because the real due date

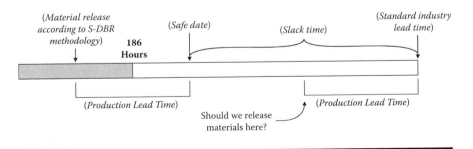

**Figure 4.4  What to do with slack time.**

was established 30 days from now and the production buffer is only 10 days, why shouldn't we release the order 20 days from now? This is a good, commonsense decision, but it carries one small negative ramification. Delaying the release could result in idle time at the CCR, even though we now have a firm order in hand that could reach the CCR on time (much earlier than required, in fact).

Considering the current queue of orders, releasing the order three days from now ensures full utilization of the CCR without overloading the shop floor. If demand continues to be low, the CCR will certainly be idle for a significant amount of time. But if the demand increases in the very near future, being able to better utilize the CCR will pay dividends.

When we release the order based exclusively on the planned load and the time to actual promised delivery is longer than the required production buffer, *de facto* we have extended the production buffer for that order. The extra time added to the buffer is called *slack*. That slack isn't really needed, but it could relieve pressure if it turns out that we must subsequently resolve more than a few red orders. We need not expedite that order in order to finish it before the calculated safe date. Rather, we could take our time as long as we don't approach the real due date too closely.

The *slack* defined above is limited by the standard lead time. While it's possible to have orders with due dates farther into the future, we want to avoid including them in the planned load, and we certainly won't give them too much slack. Thus, the definition of the standard lead time and how it impacts the horizon of the planning are important issues. Let's deal first with the impact of capacity on customer response time. For that, we'll examine the required planning horizon.

## Monitoring Capacity: Increase It or Not?

Do we have enough capacity? How do we know? This is an important question for any organization to be able to answer. When that answer is clearly positive, it means that it's possible to satisfy more market demand than the organization is currently experiencing. When will the organization approach the limit of satisfying the demand? If we consider being reliable and satisfying the market as critically important, identifying ahead of time that the organization is closing in on its capacity limits becomes very important.

Suppose a certain resource has so much work waiting for it that it would take seven working days to do it all, and more work is continually arriving. Is this a problem? Is the resource a bottleneck (meaning the load placed on it is greater than its capacity)?

Before we can answer these questions, we need to know whether a delay of up to seven days would cause a particular order to be delivered late to the customer, or whether it would force us to quote a long delivery time, which the customer also might not like. After all, we know that some of the work awaiting resource availability will be completed in seven days or sooner.

To assess the state of capacity, we must know the "tolerance" of the market. In other words, what response time do customers expect, from the time they place the order until they receive it? Ultimately, we require enough capacity to respond within the time that customers expect. In manufacturing, it's often possible to make to stock. Doing so can enable a much faster response to demand than can be achieved by starting to work from scratch. In fact this is the most common solution when customers demand faster response than a make-to-order process can deliver. For now we'll concentrate strictly on make-to-order, but later in Chapter 6 we will focus on make-to-stock issues.

To understand the huge impact of the tolerance time of the customers, let's use a gas station as an example. What's the capacity utilization of most gas pumps?* In most cases, a gas station with only one pump usually has enough capacity to serve all the customers it's likely to see in a day. But how long are *you* prepared to wait in the queue for one gas pump? Certainly less than an hour; maybe less than five minutes. Let's assume that the maximum time some fairly tolerant customers are prepared to wait is 10 minutes. This is critical information, and it makes a real difference. If your gas station has only one pump and is located in a populated area that makes waiting time longer than 10 minutes at certain times in the day (when many customers find it convenient to fuel their cars), it will lose customers to other stations nearby that can provide pump access in 10 minutes or less, just because sometimes the waiting time is longer.

Now, the tolerance time of a manufacturer's customers is not a uniform or constant number. Some customers have shorter tolerance than others. And one reliable supplier could see more tolerance from the market than the same customer gives to another less reliable supplier. Moreover, customers never tell us what their real delivery tolerance is. If they want faster delivery, why would they share their maximum tolerance time? Nevertheless, just because we may not know exactly what it is doesn't mean that customer tolerance time can be ignored. Because it's critical to the question of when to increase capacity, it must be considered.

## *Primary Time Horizon*

Thus, the term *standard lead time* really must be reevaluated. When the market as a whole has an informal standard for response time, there is a definite relationship to the tolerance time, though the two aren't necessarily equal. A customer's tolerance time might be somewhat longer because clients don't always rely on perfect due date performance from suppliers. A very reliable supplier might prompt a customer to extend the standard to the limits of the tolerance time. It's imperative that companies determine a reasonable tolerance time because offering longer delivery times is extremely risky. Assuming our product is substitutable by a competitor, we could

---

* Let's assume there are no gasoline shortages—supply is not an issue.

lose business. Let's call the assessment of reasonable customer tolerance time our Primary Time Horizon (PTH).

The reason it's called "primary" is that the planned load should be related to that time. This means that all regular orders to be delivered within the PTH should be considered in the regular planned load. Orders that lie beyond the PTH are not considered in the planned load, so they also don't have assigned release dates. This means the slack—the additional buffer resulting from early release of material because of a small planned load—is limited by the size of the PTH. For instance, if the PTH is four weeks and a customer asks for a delivery five weeks from now, that order will not be part of the current planned load. No slack is given to it at that time. A week later, the due date of that order enters PTH, and it is reflected in the planned load for delivery within the PTH. In this case, the order could have been assigned slack if the planned load is relatively small.

The resulting safe date from the planned load should never be allowed to approach PTH too closely. The PTH is the line where real damage might be caused to the organization. The planned load should consume no more than half of the buffer ahead of the PTH. If several buffer sizes are in use for different product lines, consider the "half" criterion to apply to the longest of them.

Any decision to increase capacity should account for the rate the planned load is likely to grow (sales increasing at a reasonable pace). Besides PTH, the other key parameter is the time it takes to acquire more capacity. The decision needs to be made when calculation shows that, assuming the pace of growth continues, the planned load will consume the full PTH in the time required to put into place the contemplated capacity increase.

## What Capacity Should Be Increased?

Obviously, the first consideration for a capacity increase is the CCR—the resource with which the planned load is most concerned. But is it the only candidate? Increasing capacity at the CCR constitutes *elevation*, as described in the Five Focusing Steps (Chapter 3, Figure 3.4). When the CCR is elevated with all other resource capacity remaining the same, it's highly probable that the production constraint will shift to another resource.

Is it permissible to allow the CCR to move to another resource? The obvious answer is yes. If the new CCR currently has adequate capacity to deal with the growing sales and still satisfy the market, then "no harm, no foul."

However, there are a couple of other considerations here.

- Will the organization need to change any basic procedures once the CCR moves to another resource? Is there a threat to the stability of the organization when such a drastic change occurs?
- Will the CCR be easier to control or manage at the new location than at the old one? Or is the difference irrelevant?

If we were using the traditional DBR approach, the required changes implied by the shift of a CCR to a new location would be considerable. But with S-DBR, this is not necessarily the case. Because S-DBR assumes the market demand is *always* the major constraint, even the CCR has to subordinate to the commitments we make to the market.

S-DBR does not treat the planned load as a schedule that must be closely followed. Rather, in the execution phase it provides the CCR latitude to observe actual priorities, as determined by buffer management. Thus, no strict subordination to the CCR is required anywhere else, so the impact on procedures should be negligible. Efforts at all resources unite in achieving subordination to our commitments to the market. If the CCR moves to another resource, all that is required is to use the planned load of the new CCR as the pacing mechanism.

A CCR move will likely impact the quotation of subsequent orders. Assuming we realized an incremental increase in capacity after the former CCR was elevated, all subsequent quotations would have earlier safe dates. There is nothing wrong with that; in fact, it's highly desirable. However, nothing else would change. The procedures remain exactly the same, only the focus of attention is different—the planned load on a new resource. What could be simpler than that?

We are not yet finished with the question of what capacity should be increased. Our real concerns are not the fact that the CCR might move to another resource. A much more insidious concern is the loss of substantial excess capacity on some number of nonconstraints. There is a real possibility that we would end up with interactive constraints. This is, of course, not a desirable outcome. Besides the challenge of trying to manage "wandering bottlenecks," it's the excess capacity at nonconstraints that gives us the capability to respond—to catch up—when production experiences physical interruptions in flow or minor spikes in demand that might temporarily increase the pressure on the system.

Therefore, when a CCR is elevated, several other resources may require elevation as well so as to preserve enough protective capacity around the CCR (the new one—or the old one, if we decide to keep it where it was). This, of course, makes the investment decision both more critical and, perhaps, more complicated. But assuming sales are increasing, the decision should more than merely offset the investment.

## Capacity Buffers

One of the major insights TOC provides is the realization that a capacity buffer is often required to ensure acceptable delivery performance to the market, and the size of the capacity buffer required is sometimes fairly significant. The term *capacity buffer* has often been used in a multiproject management environment to compensate for variation in the time a critical resource would be available for another project. For manufacturing, its purpose is to establish a level of extra capacity at the CCR and other heavily loaded resources to protect the organization from a sudden surge in sales, without having to disappoint some customers by delaying their order fulfillment. Capacity buffers in a production environment are referred to as *protective capacity*. But what about additional capacity that might be acquired very quickly, such as overtime? Can that be part of the protective capacity?

The answer is definitely yes. Any available capacity should be considered part of total capacity, even if the use of that capacity incurs additional expenses. The only important consideration should be that *when you need capacity, it's available to be used*. For instance, this additional capacity is not merely theoretical or conditional when the approval of the chief financial officer or controller may be required for every hour of overtime. In order to consider such capacity as an integral part of the required protective capacity, unequivocal authority is required to be able to use it whenever truly necessary. When several orders are "in the red," the production manager must have the authority to apply overtime or a second shift for all the resources required to accelerate red orders through the production line.

But this type of protective capacity is *not part of normal internal capacity*. Rather, it is capacity that can be activated quickly in abnormal circumstances. Such use does incur some extra expense, which, of course, increases operating expense (OE). External protective capacity can be very useful because it truly protects against fluctuations in demand without requiring the huge investment of, say, a plant expansion. Overtime, extra shifts, and outsourcing represent effective ways to temporarily increase capacity without an irrevocable decision to increase facility size.

Deciding how much protective capacity we should include requires us to consider the dilemma faced if we don't have that capacity. Figure 4.5 identifies some of the factors and assumptions that often enter the decision process regarding whether to include protective capacity. The most effective way is to consider the protective capacity exclusively as capacity that should be there if there is a sudden need for spare capacity. Its expense is somewhat moderated because we have little need to add extra capacity much beyond the CCR. From scanning the assumptions, you can evaluate how valid they are in your situation and make an informed decision on a reasonable level of protective capacity. While there is a cost for protective capacity, there may be an even greater potential for loss if we decide not to include it.

Having a capacity buffer as a contingency—reserving some extra capacity at the cost of an incremental increase in expenses—is very effective when there is a relatively high probability that urgent orders will come in, and we feel obliged to deliver them quickly (possibly at a very favorable price markup). It is similarly effective in transition periods, when demand grows and steps have been taken to increase internal capacity, but they haven't yet taken effect. If the capacity increase is not fully matched with the actual increase in demand, it can simplify the management problem. Access to this kind of surge capacity eases growing pains, yet at the same time helps stability.

It must be noted that planned load calculations do *not* normally take into account those extra capacity buffers, unless there is a careful decision to include them as part of the planning and to use the additional available capacity to support

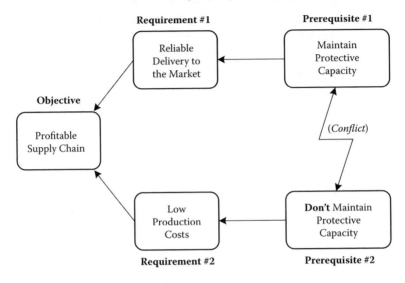

**Assumptions:**
1. Protective capacity is needed to cope with uncertainty
2. Uncertainty is a part of life in the supply chain
3. We may not be able to expand capacity when needed
4. Market opportunities may require quick response
5. Exact capacity requirements are uncertain

**Assumptions:**
6. Excess (protective) capacity wastes money
7. We should fill excess capacity with new orders
8. We are able to plan capacity needs
9. Customers won't go elsewhere if we miss due dates
10. Low cost production is the most profitable

**Figure 4.5 Protective capacity dilemma.**

increasing sales. In other words, as a normal practice we don't promise delivery dates based on this "emergency" capacity. To do so would be to compromise its value as a safety mechanism. Warning: *If you do include protective capacity in normal planning, you must be very careful to provide other protective capacity.* Otherwise, your "insurance" is already committed to the capacity required to meet known commitments, and "protection" is no longer available.

## *Performance Measures for Reliability*

How should we measure reliability? The first parameter is, of course, on-time delivery. Another is quality. Both parameters are important, and so is the choice of measurements for each. There is no shortage of references on quality measurement and management. For our purposes, we'll confine the discussion to the on-time part, at which TOC particularly excels. Regardless of the measurements selected, here are some common issues relating to on-time delivery reliability:

- What is the appropriate due date?
  - The one first requested by the customer, or
  - The one finally agreed upon by both the supplier and the customer?
- How is "completion" defined at that time?
  - Production declares the order complete
  - The order has been physically shipped to the customer, or
  - Safe order arrival at the customer's chosen location.
- Who is responsible to accomplish accurate reporting concerning due dates and completion?

These issues must be clearly resolved in a detailed procedure, but we won't address them further here. The point is that they must not be overlooked.

The most simple, straightforward measurement of on-time delivery is the percentage of the orders shipped on time.* There are two main drawbacks in using this as a measure:

1. On-time delivery is only a "yes or no" metric. If an order is late, it doesn't address the question of how late.
2. The size of the order and its financial value are not considered at all.

The conundrum associated with any kind of performance measurement is whether it might cause certain behaviors (presumably unintended) that are detrimental to the organization. Let's assume, for instance, that the accepted due date performance measurement is the percentage of orders completed on time. If this is the case, if an order turns out to be late, there's no real incentive to hurry to complete it. Once the "metric damage" is done, there is no further motivation to try to minimize the damage of late delivery to the customer.

The second concern is the size of the order and its financial value. What if two or more orders compete for the CCR, or some other resource that happens to be fully loaded at that time, and it's possible to expedite only one order to salvage an on-time delivery? Which order would we want it to be? Because we would prefer to minimize the damage to both our customer and our reputation, the financial value of both orders and the resulting delay of each should be considered.

Let's address the first problem. Any performance reliability measure should consider the amount of time that the order is late. One idea might be to tally daily

---

* Delivery due dates must be adjusted for reliability of shipping methods. Transportation buffers must be included in planning shipping dates to ensure actual delivery meets customer requirements.

**Table 4.2  Days Late by Order (for the month)**

**Day of the Month**

| Order | 1 | 2 | 3 | 4 | 5 | 6 | | 30 | 31 | Total |
|---|---|---|---|---|---|---|---|---|---|---|
| 1 | 1 | 1 | 1 | 0 | 0 | 0 | | 0 | 0 | 3 |
| 2 | 1 | 0 | 0 | 0 | 0 | 0 | | 0 | 0 | 1 |
| 3 | | — | 1 | 0 | 0 | 0 | | 0 | 0 | 1 |
| 4 | | | | | 0 | 0 | | 0 | 0 | 0 |
| 5 | | | | | | 0 | | 0 | 0 | 0 |
| | | | | | | | | | | |
| | | | | | | | | | | |
| 99 | | | | | | | | 0 | 0 | 0 |
| 100 | | | | | | | | | 0 | 0 |
| | | | | | | | | Total Late-Days Counter | | 5 |

the number of already-late open orders and multiply each one by the number of days it is late. The result is a metric we might call *late-order days.*

For example, let's say that on the first of the month, we have two orders that should have been shipped, but are late. So today the "late-order days counter" reads two (refer to Table 4.2.). The next day, one order is shipped, but the other is still late. The counter now reflects three late-order days. (One for the order that is only one day late in shipping, plus two more for the order that is now into its second day late). The third day the original order from the first day is still late, and we have failed to ship another order that was due to go out on day three. We now have five late-order days—the three carried over from yesterday, another from the old order, plus one more from a new order.

On the fourth day, the two late orders from yesterday are shipped. No new late orders occur, so the counter still tallies five late-order days. Let's assume that all the remaining orders for the month are shipped on time. So the on-time measurement of late-order days is five. Is that bad? Well, how many orders did you ship this month? Suppose 100 orders were shipped, so the percentage of the counter to the number of orders supplied is 5 percent.

Thus, one possible measurement might be *the ratio of the "late-order days monthly counter" to the number of orders shipped that month, expressed as a percentage.*

However, there are potential problems with this metric that must be addressed. For example, what if no orders are shipped during a given month? A zero value would look like "no late orders," rather than no measurement at all. What about a situation in which the measurement comes out over 100 percent? Suppose, for example, we had only one order to ship, but we weren't able to ship it until the 30th. In this case, the counter would read "30." With just the one order as the divisor, our monthly reliability measurement seems to be 3,000 percent for that month. This is certainly a highly unfavorable measurement, and though it makes sense mathematically, it doesn't really paint an accurate reality picture.

Let's put these inconvenient details aside for the moment. A good software algorithm could help us achieve the same objective we're trying to achieve—taking into account the degree to which an order was late as well as the number of orders—without distortion.

## Incorporating Financial Value into Performance Reliability

Instead, let's turn our attention to the size of the late order or its financial value. This important element is missing in the late-order days measurement. One order might have a very different financial value than another. A single order for 1,000 units selling for $90 each has a decidedly different impact on revenue than does a single order for 10 units that sell for $150 each. Equating the two in importance is likely to be poor judgment, and it would severely skew our measurement system.

Somehow the financial value of the order must be incorporated in the late-order day measure. And in actuality, we don't even really have to count orders. Rather, we can combine the money value of the order with how many days it's late.

Let's review the preceding example, but now we will introduce the financial value of the orders (refer to Table 4.3).

On the first of the month, we had two orders that should have been shipped, but were late. Order 1 is worth $10,000. Order 2 is worth $500. Therefore, the total of late-order value days is $10,500. The next day, the smaller order is shipped, but the larger one is still late. The counter now increases by $10,000 to $20,500.

On the third day, Order 1 is still late, and we fail to ship Order 3, due that day, worth another $12,000. Our late-order counter is now $42,500. On the fourth day, the two late orders from the third day were shipped, and no new orders are late. The counter remains $42,500. But this is not "real money"; it's the order dollar-value times the number of days the order is late. In other words, it's dollar days.

Notice that if we had shipped Order 1 on the second day instead of the fourth day, we would save 20,000 dollar days. Even if the other orders were still as late, as reflected in the table, the counter would only total $22,500. There are two lessons here. The first is that the choice of which order to expedite can have a significant impact on the monthly measurement value. The second is that a financial value-time metric can point to obvious choices for expediting decisions.

Now, while we consider the benefit of having a measurement to support decisions regarding expediting, it is only fair to note that such a measurement still

**Table 4.3  Dollar-value Days**

| Order | 1 | 2 | 3 | 4 | 5 | 6 | | 30 | 31 | Total |
|---|---|---|---|---|---|---|---|---|---|---|
| | | | **Day of the Month** | | | | | | | |
| 1 | $10,000 | $10,000 | $10,000 | 0 | 0 | 0 | | 0 | 0 | $30,000 |
| 2 | $500 | 0 | 0 | 0 | 0 | 0 | | 0 | 0 | $500 |
| 3 | | — | $12,000 | 0 | 0 | 0 | | 0 | 0 | $12,000 |
| 4 | | | | | 0 | 0 | | 0 | 0 | 0 |
| 5 | | | | | | 0 | | 0 | 0 | 0 |
| | | | | | | | | | | |
| 99 | | | | | | | | 0 | 0 | 0 |
| 100 | | | | | | | | | 0 | 0 |
| | | | | | | | | Late-Days Counter | $42,500 | |

doesn't address some other relevant considerations. Suppose Order 2, though it has a relatively small financial value, is intended for a very important client, with whom we should maintain an excellent high level of satisfaction (significant future business may depend on it). So while dollar days give us much better information in identifying which order to expedite, beware. When it comes to a disappointed client, the value of the immediate order is not always the most significant factor.

There are two additional issues we must address. First, when we assigned a financial value to the order, what value did we refer to? Second, should we use that measure without comparing it to the total orders shipped or not?

## *What Is the Value of an Order?*

The natural tendency is to use the selling price because it represents the value to the customer. But one could argue that this is not the value of the order to the organization. For the producer, the added value of a sale is *only the throughput* of the order (selling price minus the raw material cost). Thus, for the producer, the measure of impact for a late order should be the "T" of the order times the number of days late that the order was shipped (throughput value days, TVD, or throughput dollar days, TDD).*

---

* We use the term "dollar" throughout this book in lieu of "value." For readers whose currency is something other than dollars, substitute your own currency.

However, as a reliability measurement, shouldn't our measure of merit take into account the customer's perception of damage if our delivery dates aren't reliable? A customer's perception of damage is both variable and uncertain. There's no way to measure it precisely, but we can safely assume that it would be higher than the value of our throughput. The only increased value we can possibly assign to such damage with any degree of confidence would be our selling price.* Thus, when we use the selling price in the late-order value days, we concentrate on an estimate of damage to the customer. When we use throughput, we focus on an estimate of our own damage. Does the difference truly matter?

The real value of using the late-order value days is to have a reference for improvement. Let's consider the reference of *the rate of late-order value days we now have*. Now we have a reference for next period of time: our target is to get better, to have significantly less late-order value days. Thus, in measuring relative performance for decisions concerning improvements, there is no real difference between using the selling price or throughput. The key is to use a good estimate of the value of the order and the relative lateness.

Of course, once the measurement is in place, setting the monthly graph of the monthly late-order value days makes a lot of sense and shows in detail whether the company is improving in its efforts to be truly reliable.

There's one more issue to deal with. The problem with all measurements is that people may understand the logic, but sometimes they fail to achieve good results. Sometimes this may prompt them to try to "mess with the measurement." For instance, suppose a client order consisting of 20 different products was to be shipped in a specific date, and we managed to ship 19 of them on time. How should this be reflected in the measurement—calculate the missing product as one order that was late, or consider the whole package late? It would certainly have an impact on throughput dollar days. We can be fairly certain that operations would like to claim that only one product missed the shipment. But is it really only one product? Suppose the customer really needs all 20 products. What would the customer think of our due-date performance?

This is one of the details of constructing a measurement that can significantly affect what the measurement looks like. If we split the big order into separate orders, we might consider that only 5 percent of the orders were late, and then take the financial impact of just that product. But if you *don't* split the order, *and* you *do use* the financial value of the order, the whole order must be considered late. The impact on the resulting measurement is significant.

This issue—how to model aggregation of items into one order—is a key to understanding the difference between the simple measurement of *late-order days*

---

* In reality, our late delivery could potentially cause customers damage to their own reputation, lost potential future sales, etc. But we have no way of assigning an accurate value to those losses, much less measuring them. So we must be content with an arbitrary metric, and selling price is about the only one for which we can assign a number.

and the much more sophisticated one of *late-order value days*. The issue is whether it is truly necessary to reflect the financial value of the orders. On one hand, using *late-order value days* would direct the shop floor to concentrate on the higher-value orders, and this would minimize the damage to customers. On the other hand, is it really true that the higher value the order, the higher the damage of lateness? And what about the customers themselves? Shouldn't we be more careful about a big customer?

Buffer management offers priorities that have nothing to do with the financial worth of any order. The idea is that, following the buffer management priorities, all orders would be on time. Only when that assumption (all the orders will be completed on time) is clearly not valid does the question of which order we cannot afford to deliver late comes up. And the financial worth of the order is only one parameter in that decision.

We consider *late-order value days* to be an excellent periodic measurement of reliability, but it should not be used on daily basis where buffer management is the only robust priority system. It should be used only for management's periodic analysis of the state of the company and how well it's doing in providing value to its customers.

In most cases, the simpler *late-order days* is adequate, and its simplicity is a real asset. From the information system point of view, it doesn't require consideration of the selling price of the order. Moreover, it also conforms better to intuition. So in most cases, we would suggest choosing this as the main reliability measure.

While TOC does not deal with quality measurements, one aspect of quality has to be clarified: if an order is returned by the customer, then that order was NOT delivered on time. And the days late should include all the days between the original due date and the final delivery of the order, approved by the customer.

It's time to now examine some of the unusual circumstances that have surfaced in the application of S-DBR over the past eight years.

*Chapter 5*

# Overcoming Complications in Using S-DBR

## Contents

Simplicity is the most striking characteristic of simplified drum–buffer–rope (S-DBR). We consider simple solutions a huge advantage because only a truly simple solution has a high probability of working effectively consistently. But at the same time, it's natural to question whether S-DBR isn't just too simple. Are there limitations to S-DBR, as described so far, in handling really complex production processes? It's time to examine several complicating factors and determine whether these complexities might prevent a successful S-DBR implementation.* The topics we'll examine in this chapter include:

- Multiple operations of the capacity-constrained resource (CCR) on the same order
- Multiple simultaneous CCRs
- "Wandering" bottlenecks
- Manpower as a CCR
- Long manufacturing process times (including subcontracting)
- Exceptionally large orders
- Sequence-dependent setups

## Multiple Operations of the CCR

There are two distinct situations in which a resource must perform multiple operations (in some sequence) on the same order. Before we consider these situations, let's be clear that if such a resource has enough protective capacity to be considered a nonconstraint, S-DBR ignores it. It's assumed that a nonconstrained resource can find the time, within the overall time buffer, to do whatever might be required. Consequently, in this discussion most of our emphasis will be on situations in which a CCR must perform multiple operations on the same product. If we can conclude that S-DBR accommodates this particular situation, it should be clear that S-DBR will have no problem in managing other nonconstrained resources with multiple operations.

The first situation is an assembly of several components where some or all require the same machine. Let's use an "A" plant to examine the situation in which multiple CCR operations pertain in more detail. Figure 5.1 shows three major components in parallel paths that require machine #1 (M1) for a specific operation.

Another multiple CCR configuration might be sequential (Figure 5.2). In this case, the CCR is required more than once in the same linear sequence. This situation is often referred to as a "reentrant" environment. The same work order returns to the same resource (or is multiple resourced) more than once.

---

* These complications are no less relevant in traditional DBR, lean manufacturing, or even manufacturing resource planning (MRP), nor are they any easier to accommodate in any of these other methods.

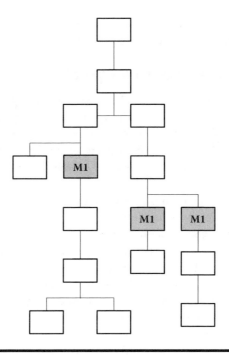

**Figure 5.1    Three CCR operations in an "A" plant.**

Generally, the latter configuration is considered more complicated than that indicated in Figure 5.1 because M1 is required to go through the operations in a sequence. When there are intermediate operations that must be accomplished by other resources, M1 has either to wait for the order to return again or process other orders between the different operations for the original order.

In S-DBR, by managing the planned load neither of these situations poses any real difficulty. The production buffer still protects the entire routing, whether it contains assembly of many components, a reentrant sequence, or even a combination of the two. Naturally, a very complicated routing could require a longer production buffer, but this depends whether the additional complexity increases variability (that is, Murphy's law*). The number of setups, and possibly the longer processing time at the CCR, usually increases the production buffer only slightly. Normally, the production buffer is adequately sized to begin with. Providing a sufficiently sized buffer leaves the operators to handle the complexity by deciding among themselves on the sequence M1 should go through. Buffer management is very important in this instance. It must guide when the operator should turn his/her attention to conclude some orders for which the finishing M1 operation is

---

* "Whatever *can* go wrong, *will* go wrong, and it will do so when least expected and most inconvient."

required urgently in order to be on time, and when it is safe to introduce some of a newer order into the upstream operations of M1.

## *Planned Load in Multiple CCR Operations*

What might need some clarification is the treatment of the planned load. The entering argument for the planned load is the CCR capacity required for an order that passes through the CCR multiple times. However, only one summation of capacity is required per order. So when calculating planned load, the input data must consider the total time required by the CCR for that order—the sum of each pass through the CCR.

In certain cases, especially in the reentrant category, that capacity is not going to be expended on the order continuously because there is no assurance that the multiple operations will be done consecutively. But the timing calculated in the planned load is continuous, and based on this timing the due date is determined as well as the release date. This means the actual time the CCR will work on the orders cannot be the same as used in the planned load. Does this mean that we did something wrong by determining the safe date and the material release based on the planned load?

Not really. Remember that *the planned load is not a schedule*, thus the requirement for more than one operation at the CCR is not an issue. In reality, CCR operations are not necessarily processed at the exact time and sequence indicated in the planned load. When a number of orders have multiple CCR sequential operations, the CCR is likely to finish the first operation for order X, then shift to a required operation for order Y because the parts for the second operation on order X are not yet queued at the CCR. Only after completing the first operation on order Y would the CCR revert to processing order X again in the second operation.

Suppose we have an order, whose routing is shown by Figure 5.2, and the planned load is now at 240 hours; all M1 operations together will take 8 hours of M1; and the production buffer is 80 hours. We plan, according to the last chapter, for the safe date to be 280 hours from now, and we intend to release the materials 200 hours from now. But since the order requires four different M1 operations, there is no chance all the four operations would be processed by 240 hours from now. It could be that the first M1 operation will be done at the 240th hour, will take 2 hours, but then need to go through two different operations before returning to M1. So the second M1 operation could be, say, at 260 hours, then again at 268 and at 276 hours (the order is actually in red after hour 256). Starting the various CCR operations in the middle of the time buffer

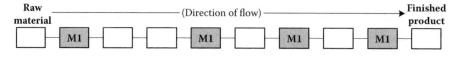

**Figure 5.2   Multiple CCR operations: Sequential.**

is an exceptional case. In a more normal case, the first M1 operation could be assumed to be completed at about 220 hours, the next M1 operation at 235, then at 246 and 256. These successive interations are all roughly around 240, but not at 240 *at* the four operations.

Therefore, when the planned load releases an order that includes several CCR operations, half the length of the buffer before its assumed time, in most cases the first operation would be done *before that time*, and the last operation after that time. The production buffer has to give the CCR operator enough of a choice so as not to waste capacity when the pressure is on, yet not permit too much choice. If we must provide many choices for the CCR, then the way to do it is to increase the production buffer. Unfortunately, this adds more work in process (WIP), and this would cause too long a wait time, especially for the CCR. All in all, constricting the release of material so the number of orders in the floor is limited, and practicing effective buffer management priorities, provides a stable way to be both agile and reliable even at what looks like a complicated environment. Actually it is not complicated at all, as long as the CCR has adequate capacity and all the rest of the resources have even more capacity.

## Multiple Capacity-Constrained Resources (CCRs)

It is entirely possible to have two or more work centers with approximately equal capacity and those work centers are the "weakest links" of the shop floor. This is *not* a desirable situation, and we recommend taking steps to "unequalize" the capacity. But if this isn't possible, or if operations must continue for an extended period until the capacity can be unbalanced, this is how to manage production.

In essence, this case is a situation in which one CCR feeds another. Most managers think it ideal to have all resources—people, machines, tools, etc.—fully loaded to 100 percent. More realistic executives might want to have all resources loaded to 90 percent of capacity because they *do* understand the need for some protective capacity, but they would still strive to increase efficiency by maintaining very little spare capacity. While just-in-time (JIT) implementations that produce standard products usually have 8 to 10 percent protective capacity, we recognize that more will be required for all other circumstances.

What even those executives in JIT operations fail to recognize is that *having a balanced shop floor means operating in a chaotic situation*. One of the resulting effects is losing the ability to focus, which strengthens the feeling of being in a chaotic environment. A balanced line is subject to variability. Each time some variation or interruption occurs—and these inevitably happen randomly within the process—it precipitates an ongoing search to determine what went wrong. Today it's one thing at one location; tomorrow it's another somewhere else. Worse, the variations of closely balanced resources accumulate, since variation in one resource also impacts succeeding ones. The net result is disruption of deliveries to the market.

Management spends its time reacting to problems rather than proactively managing the process. They don't recognize that a 90 percent capacity loading is too high, where "too high" implies that there is no maneuvering room to deal with something that they can be certain is going to go wrong.

Interestingly enough, there's one case where having two or more CCRs poses no problem at all. If, under one roof, you run two or more families of products that use dedicated machines, then having one CCR per family is quite acceptable. In effect, you are simply running two or more different plants in the same facility. Applying S-DBR in each independent product line would not impose any complication.

Complications arise when the two or more CCRs are interactive in the same line, which means the output of one machine enters the other as input (even if there are intervening steps). Managing the interactions when both resources have very limited protective capacity usually turns out to be a real headache. Organizations pay a very high price for their ignorance about such a situation—or their failure to do anything about it, if they *are* aware of it.*

So what should we do when we have more than one CCR especially if our options for overtime, extra shifts, or outsourcing are limited?

You can try to complicate the basic S-DBR methodology, but we would lay odds that you won't find a way to do so that's truly reliable. You can only avoid the inevitable strategic managerial decision for so long. Eventually you're faced with one of two choices: either you control your shop floor by having adequate protective capacity and only one focal point of your planning and control—the CCR—or you try to squeeze the last drop of blood from your capacity, with a very high chance of losing your market (not to mention developing ulcers). Obviously, we suggest the former.

There is yet another situation where more than one CCR might be active, and a somewhat different managerial decision might set it right. Figure 5.3 illustrates what it might look like.

There might be markets for partial products, and overall market demand could be very large. Steel producers face such a situation, especially when the world demand for steel is high.

Figure 5.3 presents a situation in which the CCR for the highest "T" Product (the one that generates the highest throughput) comes from either M6 or M7. Thus,

---

* If you want to experience some of the effects of having two resources of nearly equal capacity, try the Management Interactive Case Study Simulator (MICSS) scenario called *Utopia*. It's based on the ADV200 scenario (described in detail in *Manufacturing at Warp Speed*), but in Utopia the capacities of the MB and MD resources are nearly the same. Then, if you want to feel the real impact, add a second shift, but *only* when you realize you have no choice. Make it your priority not to waste money "for nothing." Remember, you have three resources that should not be idle most of time. Besides MB and MD, you have also GT to consider. The MICSS software, including the new Utopia scenario along with others, and other related materials, are available for download from two Web sites, http://www.inherentsimplicity.com and http://www.goalsys.com

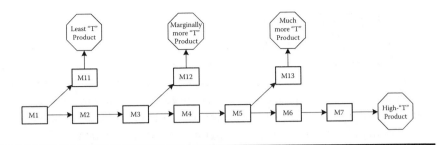

**Figure 5.3 Multiple CCRs: A special case.**

M4 and M5 have more capacity than M7 (assuming it is the CCR), but they can also supply another market—not as lucrative, but still reasonably good. Consequently, all the spare capacity of the slowest machine among M4, M5, and M13 should be dedicated to the Much More T Product. M1, M2, and M3 have to support *two* CCRs, and they, too, still have some spare capacity. However, there is another market for a product we call Marginally More T. And M1 must squeeze every bit of its capacity to supply everything M2 needs. If any capacity at all remains, it would be directed toward the Least T Product.

Figure 5.3 shows four interactive CCRs. Let's assume the market for the High T Product demands our maximum efforts for due date reliability because it returns the best throughput. How can the other three CCRs (M1, M3, and M5) fully support M6 and M7, while they're striving to extract the most output from their capacity?

Strategically, the preferred management decision would be to stringently limit efforts to capitalize on the less lucrative markets if there is even a remote chance of failing to be fully reliable for the best markets. The global priority should be clear.

The Highest T market should be 100 percent secure. Only then could M5 squeeze an order for the Much More T market. Acceptance of such an order by M5 should be based on its own capacity analysis, after prioritizing all the orders for the Highest T Products, and only when the planned load for these orders allows. Only then would M5 be permitted to take additional orders for the Much More T Product. The same logic should apply to M3. But in this case, M3 must consider all the demands of M5 as the high priority load. Only when there is clearly enough spare capacity after fulfilling all of M5's needs could producing for the Marginally More T market be considered.

To summarize, we should create very clear priority rules that will ensure protecting the more lucrative markets, even at expense of losing some sales in the less lucrative markets. This way, even though we technically have several CCRs, our clear hierarchy helps us to resolve the contention among them. Obviously, this is a somewhat complicated solution. In such implementations, we must be very careful to follow the throughput-based sequence priority.

## Wandering Bottlenecks

"Our bottleneck moves practically every day." This is a common refrain from operations managers. It's the most common objection to DBR. But such a statement actually substantiates the reality that the production shop floor is in a chaotic state, which usually forces the production manager to run from pillar to post, nonstop, trying to "put out fires." Two different possible causes might result in bottlenecks occurring all the time. Both causes misinterpret the actual situation.

*Policies Drive Behavior*—The first erroneous perception stems from flawed policies that drive normal behavior. Chief among these is striving for "efficiency" as a mean of achieving lower operating expenses. Let's see what happens when large process batches are used and transfer batches are naturally equalized to the process batch.*

Suppose a large batch is released to the floor. In this case, "large" means that it might take several days for one or two work centers to process all of it. When this batch reaches the first work center, the load on that center goes up considerably. If we assume normal human behavior†, the operator of this work center likely has already adjusted his pace to ensure that the work center appears to be fully occupied with work all of the time. Consequently, this operator isn't idle. Preceding work orders are likely to be still at the work center.

Now the new large order arrives and dramatically increases the total WIP queued at the work center. Any subsequent work order that arrives will wait for all the previously accumulated work to pass through the work center. This is what seems to be the bottleneck of the day: the work center with the largest pile of inventory.

Finally, the work center gets to the large order. Completed parts of that order wait for transfer to the next work center because the entire batch has not been finished. More time passes until the large order moves to the next work center. When this happens, the next work center now holds the largest unprocessed load among all work centers—the bottleneck has "moved"… apparently.

Adding to this scenario is the tendency to release orders as early as possible, depending on the capacity of the gating work centers (those work centers at the upstream of production); the amount of WIP in the shop becomes huge. With the transfer batch still equal to the process batch, and the fact that each work center wants to ensure that it has some work to do tomorrow, the distribution of the WIP throughout the production floor is completely arbitrary. Thus, one day a particular work center has the largest load, the next day it will be another.

*Product Mix Change*—A second cause of apparently moving bottlenecks can be an actual change in product mix. When a certain family of products dominates the product mix, a certain machine is clearly the weakest link. When the mix changes, another product family might become dominant. If this new product family makes

---

* This is not an uncommon situation in many manufacturing operations.
† Parkinson's law: "Work expands to consume the time allowed for it."

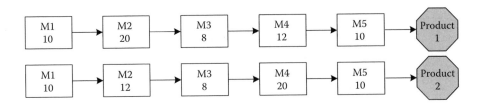

**Figure 5.4  Constraint shift from product mix changes.**

heavy use of a different work center (or machine), that work center (or machine) may emerge as a new constraint. Here is a simple demonstration (Figure 5.4).

Figure 5.4 doesn't indicate the market demand for the two products. However, if we assume that the shop produces just those two, there are two candidates for the CCR: either M2 or M4, depending on the production ratio between the two products. If considerably more of Product 1 is sold, M2 would clearly be the CCR. Conversely, if Product 2 sells more, then M4 is obviously the weakest link.

Notice again that we continually refer to the weakest link as the CCR, even though the weakest link is not necessarily an active constraint. It's convenient to use CCR to designate the one point in the shop floor that *might* become an active constraint. Knowing this enables careful monitoring of the capacity of the CCR, using the planned load to signal when the load on the CCR approaches its capacity limits.

What happens when the demand for both Product 1 and Product 2 are approximately the same? As long as both work centers experience adequate excess capacity, work flow is smooth and fast, and the only constraint is market demand. However, because of the interaction between the two work centers, there is a potential for trouble as excess capacity diminishes. This will happen at a capacity loading noticeably lower than in a situation where there is only a single, well-defined CCR. In the latter case, the protective capacity is sufficient to ensure smooth performance. But the interaction of two approximately equal work centers requires a higher level of protective capacity to ensure delivery reliability.

When two CCRs interact with each other, to an observer it appears as if the bottleneck cycles back and forth between the two. Almost any new incoming work order affects the relative load between the two. Because of the close proximity of the capacity of the two resources, their variability ranges overlap. Thus, daily variability in demand shifts the sensitive balance between the two (Figure 5.5). The real problem is the sensitive balance that causes inability to be truly reliable. When the two go through a phase where they are actually interactive constraints, the overall ability to control the reliability deteriorates as we discussed earlier. CCR shifts between the two create an interactive state without any early warning. Of course, if capacity buffers such as overtime are used, you just might get away with it. When no such rapid capacity additions are possible, one of them must be fixed as the main CCR, while carefully monitoring to ensure that the other one won't experience too high

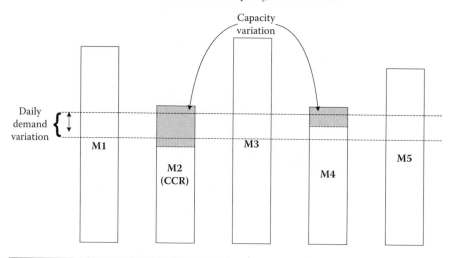

**Figure 5.5 Interactive capacity-constrained resources (CCRs).**

a demand. This means monitoring both through their respective planned loads, ensuring that the non-CCR resource will maintain a larger difference between its planned load and the primary horizon discussed in the last chapter.

For instance, if the ratio between units sold of Product 1 and Product 2 is 95-to-100, then Product 2 sells, on average, somewhat better than Product 1.

Thus, overall M4 has more load than M2. But maybe next week M2 might experience a little more load than M4. Because the loads on M2 and M4 are normally close to one another, any variation may result in one resource being more heavily loaded on a given day, and that situation might reverse itself shortly thereafter.

There's another potential complicating factor at work, too. Capacity can vary from day to day as well. Machines require maintenance, both preventive and reactive. Though the range of capacity variation for a particular resource may be small, it's nonetheless real. Even if demand remains constant, relatively small variations in the capacities of two nearly equal work centers can produce a CCR shift. When the variation in demand is overlaid on top of capacity variation, the problem is exacerbated—and worse, unpredictable.

If M4 wastes potential production capacity due to starvation, the net result could be compromised delivery reliability to customers. Without adjusting the capacity of either M2 or M4, one way to ensure that M2 is a nonconstraint is to significantly increase the production buffer. When this is done, however, the additional capacity at M2 is usually translated into more WIP between M2 and M4. Certainly, incidents of precious lost capacity at M4 are reduced. But will customers tolerate a slower response resulting from a longer time buffer, or are they likely to take their business elsewhere? Trading off time buffer against the cost of added capacity at M2 is certainly a viable option, but in the long run it may not be the best choice. After all, aren't we using S-DBR for the express purpose of ensuring higher delivery reliability in *shorter* periods of time? Deliberately "unbalancing" nearly equal capacities of two interactive CCRs, if possible, would be preferable. A production process with only one definable CCR is usually faster and more reliable than one with two or more.

## Summarizing Interactive CCRs

What can we learn from this analysis? Whether a resource is a constraint depends on the time horizon in which a response is expected. In a gas station, where the expected response time for a car entering and leaving the station is less than 30 minutes, a 15 percent average utilization of the gas pumps could be already problematic. When the expected response time of a distribution chain for towels is six weeks, a utilization of 90 percent of the weakest link might still provide good reliability, assuming the next highest loaded work center averages at least 15 percent excess capacity.

Bottlenecks may move, but they shouldn't move too quickly. If every new order shifts the constraint between resources, such as M2 and M4, be aware that both are probably interactive constraints.

What's the best way to manage interactive constraints? Not by "better scheduling." Not even by more sophisticated execution control. Any such sophisticated solution would still be subject to the uncertainty of market demand and capacity variability. Constant shifts in priorities and less than perfect reliability would still result.

The only effective way to address interactive constraints is through strategic management decision; don't fall into this interactive constraint trap in the first place. Or, if you're already in it, get out of it. Solve the problem by either increasing capacity—in other words, *elevation*, which requires additional financial investment—or by restraining the less lucrative market. The former can be done by increasing capacity on all but one interactive CCR, or by "choking" the capacity of one resource to make it the only CCR. The latter can be done by manipulating market demand through changes in price or quoted lead time. Whichever strategy is selected, the result will be one unequivocal CCR and sufficient protective capacity at all non-CCRs. At this point, it would be up to marketing and sales to capitalize on the resulting superior reliability and obtain the best price from the right market for it.

## Manpower as a CCR

Here's a real "war story." During a traditional DBR implementation, a certain manufacturer experienced too many "red" zone penetrations, and even late orders. Because it was a very complicated operation, the Main Assembly was treated as the CCR. But it was not a real bottleneck. Planned load for Main Assembly seemed to be well within nominal capacity limits. Yet a growing number of orders penetrated "into the red," and even into the black*. This is obviously an indicator of an emerging bottleneck. The question is: where?

To identify the new bottleneck, the list of red orders was reviewed and their exact location in the production process was determined. Normally this procedure works just fine, and most of the orders prove to be "stuck" in front of a specific work center, but not this time. All the red orders had passed the Main Assembly, which was the assumed CCR. Following assembly, several finishing operations had to be completed, most of the time to fix some quality deficiency. The red orders were spread, without an obvious cluster, throughout all the finishing work centers. So where was the real bottleneck? Not with the machines in the finishing work centers, it turned out (Figure 5.6).

If the machines and the tools were not the bottleneck, the only other possible constrained resource is manpower. The validating evidence was that the operators who complete all the finishing operations belong to a well-defined pool of skilled manpower, and there were not enough of them to fully man all the required work centers. This revelation changed the way the routing was viewed. The machines

---

* Black means being already late. It is a kind of informal term because we never want to be late at all, so a formal name for being late has not been established. Reality says we do need to consider "black orders" from time to time.

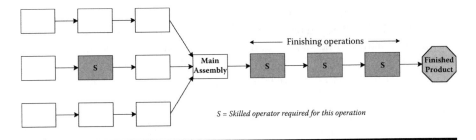

**Figure 5.6  Manpower as a capacity-constraining resource (CCR).**

were not the critical resources. In this case, the pool of skilled operators was the constraint of the organization.

A pool of operators as the capacity-constrained resource poses additional difficulties. Generally speaking, human capacity fluctuates more than does machine capacity. It becomes more challenging to assess the expected capacity utilization, so the planned load metric is less robust than when applied to equipment alone.* "Exploiting" human resources becomes a *very* sensitive issue that could fulminate into a clash between management and workers. The conundrum lies in trying to extract every quantum of capacity from one group of people, while other groups are subject to much less pressure.

Rinehart, Huxley, and Robertson (1997) describe the high-profile labor unrest that developed in the 1990s at CAMI Automotive, a unionized joint venture of General Motors and Suzuki in Canada. Unionized employees went on strike because of the excessive human stress imposed by the implementation of lean production and its emphasis on maximizing line efficiency. GM (if not Suzuki) undertook lean production with the implicit intent of paring as many jobs as possible, though they dressed it up as "improved efficiency." In doing so, they created a situation in which human resources became the CCR, with devastating results. Situations such as this require both wisdom and a thorough understanding of human psychology to achieve stable, productive, beneficial work relationships.

However, the real questions are: Why should we ever allow people to become a CCR in manufacturing in the first place? What obstacle prevents hiring and training more people? In most cases, direct labor capacity in manufacturing is relatively cheap compared with the damage a manpower constraint poses.†

Why does a pool of specialized operators so often become a CCR? The answer is that direct labor seems to be the first target for saving costs. The whole notion of efficiency is directed toward people, not toward the equipment. Companies in the

---

* This provides some justification for highly automated processes, though there is the trade-off of ensuring higher reliability–maintainability–availability of such equipment.
† And this brings us back to the elemental contention between cost-world and throughput-world thinking, a subject that is more effectively treated in greater detail elsewhere.

West have embraced lean specifically for this purpose. Thus, the cost world deliberately initiates moves that end up causing manufacturing organizations to risk losing part of their market—the "law of unintended consequences" at work. Of course, this in turn reinforces the drive to lay off more people, and a vicious circle is generated. We refer to this as a negative reinforcing loop (Figure 5.7).

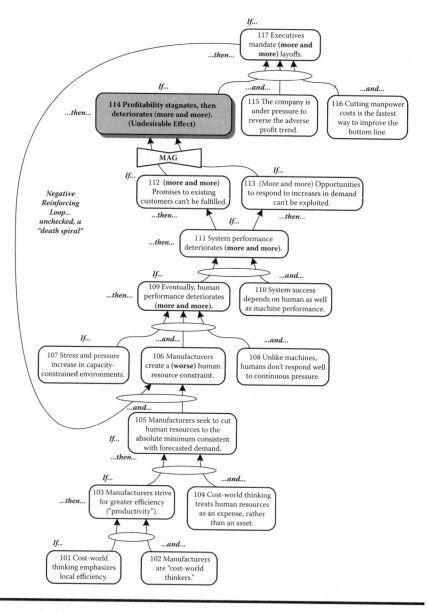

**Figure 5.7   Manpower as a CCR: A negative reinforcing loop.**

The unmistakable conclusion for manufacturing organizations should be *not to allow any part of your direct labor to be a constraint*. Don't sacrifice the capacity of a very expensive piece of equipment to the false economy of manpower cuts. By maintaining some excess capacity in your workforce, you will be able to ensure reliable promises to your customers and exploit unanticipated new opportunities.

## Long Manufacturing Process Times

A critical basic assumption underlying both DBR and S-DBR is that actual processing (touch) time in manufacturing is a very small fraction of the production lead time. In other words, most work orders wait extended periods for resources to become available for processing. Presuming that this is a valid assumption, if a resource makes a mistake in evaluating the priorities of two available work orders (in other words, the operator decided to complete order X instead of a somewhat more urgent order Y), the consequences for order Y would normally be minimal. The relatively short processing time means that such a delay doesn't consume much of the production buffer. Such a priority assessment error can be accommodated. Only red-zone orders require an absolute precedence over others because, even though the time-to-complete is relatively short, with the appropriate priority it would be sufficient for clearing the red order.

Even in S-DBR, the processing time should be no more than 20 percent of the production buffer. S-DBR relies on the flexibility given to the resources to adjust to the changing conditions imposed by Murphy's law. This already means providing a production time buffer that is considerably longer than the processing times. Moreover, in order to fully utilize the CCR and refraining from unnecessary waste of capacity, we expect the orders to arrive early to the CCR and then wait there. If the CCR does not impose significant wait time, then statistical fluctuations would also produce significant idle time at the CCR. Thus, in order to preserve effective exploitation of its capacity, we must have enough queue time at the CCR. In most manufacturing situations, net processing time is assuredly less than 10 percent of production lead time. Thus, S-DBR is able to decrease the production lead time, yet still maintain a realistic relationship between processing time and total production lead time—enough to ensure effective CCR utilization.

However, what happens when the ratio of processing time to production lead time (before installing S-DBR) exceeds 10 percent? In such a situation, it's very difficult to realize high utilization. Moreover, due date performance will likely be very low, which also means that the actual production lead time is longer than the delivery time quoted to clients. This characteristic is common in projects where every project is organized to be worked continuously, so it will be completed as early as possible. In projects, there is usually enough excess capacity at all resources to ensure that progress never stops.*

---

* TOC uses a different planning methodology—*critical chain project management*—for multiple simultaneous projects.

Does this imply that all manufacturing situations should be handled as projects when touch time is greater than 10 percent of total production lead time? Not necessarily. There are two such situations in which S-DBR would be the preferred methodology:

1. Only one particular operation in the entire process is exceptionally long, or
2. Several operations experience long but reasonably stable processing times; however all resources experience very high excess capacity. Moreover, the routing is very close to an "I"; there are no assembly operations—just a sequence of operations.

## Managing with S-DBR in the Face of One Particularly Long Operation

Let's suppose that certain large wood products need slow drying, and this requires two weeks—about *half* the total production lead time. Such an operation has two unique features: you can't accelerate it by adding capacity, and you don't need to wait for previous orders to be processed (assuming a kiln is not required for drying).

The challenge in this kind of operation is its impact on buffer management. If the production buffer is four weeks, and two weeks are consumed by normal processing, then buffer penetration is 50 percent. In traditional DBR, this would have put the production process in the yellow zone.

However, if for some reason at the 50-percent time the order is not yet at the drying operation, the odds are that it will be delivered late. Moreover, at that point nothing can be done to expedite it. The implication is that if we want to be able to promise orders reliably, we must expedite *long* before a red zone penetration occurs. We must verify that the order will arrive at the drying operation *at least* two weeks before the shipping date. We've assumed that the drying is either the final operation or very close to it. If it's not, then we should verify the arrival of the order at the drying process earlier than two weeks before the shipping date.

When the long operation is not the very first step in the process, common sense dictates protecting the part of the process from material release to the long operation with a time buffer and managing that buffer as though it were the shipping buffer.

Assuming the long operation can't be accelerated and that applying priority to an individual order won't help anything, there isn't a need to apply buffer management to it. All that's required is to allow for the long operation's lead time, which would not be part of the time buffers. If, after the long operation, completion of the order still requires additional steps, these operations should be assigned a time buffer. Figure 5.8 illustrates this configuration.

Suppose the long operation—a drying operation—takes two weeks, and that time doesn't vary much. Let's assign time buffers of a week for the normal operations before and after the long operation. The overall production lead time is a combination of the two one-week buffers, plus the two weeks required for the long operation.

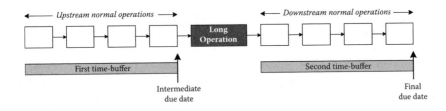

**Figure 5.8  S-DBR with a long intermediate operation.**

Where is the CCR in this example? If the long operation is not restricted by capacity, the weakest link—the resource that could become a capacity-constrained resource when the market demand is high—might be in either the upstream or at the downstream segments.

When can we confidently promise delivery to a customer for such long-operation processes? If the CCR is in the *upstream* part of the routing, then the safe date is determined as:

- Safe promise date = $L_p + \frac{1}{2}B_1 + O_L + B_2$
  (Planned load + half the upstream buffer + length of the long operation + full downstream buffer)
- Order release date = $L_p - \frac{1}{2}B_1$
  (Planned load – half the upstream buffer)

On the other hand if the CCR is in the *downstream* part of the routing, then:

- Customer Due Date = $L_p + \frac{1}{2}B_2$
  (Planned load + half the last time buffer)
- Due date for first time buffer [or $O_{L\ (START\ DATE)}$] = $L_p - \frac{1}{2}B_2 - O_L$
  (Planned load – half the last time buffer – the time for the long operation)
- Order release date = $O_{L\ (START\ DATE)} - B_1$
  (Due date for the start of the long operation – the first time buffer)

The approach described above uses multiple time buffers connected back-to-back to each other. Clearly, it's much more complicated than the fundamental, simple approach of S-DBR, and it makes response time slower. However, it's still simple enough considering the complexity of having a long time operation in the middle of the production process.

# S-DBR and Outsourcing

Sometimes parts of a process, or even a sequence of operations, are outsourced. This situation is similar to a long operation with unrestricted capacity. Most outsourcing operations consume considerable production lead time and should be regarded as long operations. The reason is that outsourcing usually involves transportation of the entire order to and from the external processing company. This, in itself, can consume considerable time.

Now, consider that most external supplier companies won't be using S-DBR in their operations. It becomes safe to assume that outsourcing operations that are part of a production process will consume a *lot* of time.

However, the one possible exception occurs when the outsourced service provider is willing to cooperate fully by agreeing to adhere to our priorities closely. The outsourced provider might or might not be using S-DBR, but once it follows our priorities, as dictated by our buffer management, it means cooperating fully. It then becomes possible for us to use S-DBR as usual, and the benefits for all concerned are considerable. There are two essential requirements for this to work:

1. We must complete every order as quickly as possible and move it on to the external supplier, which means not holding the order for a scheduled intermediate due date.
2. We and the outsourcer must adhere conscientiously to buffer management priorities.

When these requirements are met, S-DBR can be applied as usual. For collaboration with an external source organization to be effective, the financial agreement should offer rewards for fast response and for limiting red-zone orders at the outsourced location.

More often than not, however, the best we can expect is a traditional supplier business relationship with our external sources. They will promise return delivery by a certain date. We have no real expectation of having the orders returned much earlier than the quoted date, and there is always a risk of getting the order back late from external supplier. After all, the external source is doing similar work for other clients. We can't depend on them to pay much attention to an individual client's priorities—unless the external source, too, is using S-DBR. When the external supplier serves many clients, special priorities for certain clients aren't always followed even when substantial goodwill exists.

In this situation, we are under even more pressure to ensure that the order reaches the external source on time. Consequently, we really need the first time buffer to protect that date. Because we may not be able to depend on the promised date provided by the external source—we don't expect early delivery, but it might be late—we need to protect *our* final due date against both delays at the external supplier *and* delays in our final operations. The preferred model for S-DBR with outsourcing in the process is shown in Figure 5.9.

The second time buffer must ensure that even if the external supplier returns the order on time, the downstream operations don't drive the order into the red zone. Let's suppose that the external supplier promises a two-week turnaround, and the

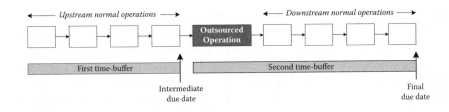

**Figure 5.9 S-DBR with an outsourced operation.**

remaining downstream operations require up to another week. Should we allow a three-week buffer for both? Doing so means that even if the external supplier delivers on time, the order is already on the verge of a red-zone penetration. Even the slightest tardiness from outsourcing would drive the order into the red. In this case, it would be prudent to add a half a week to a week to the second buffer (3½ to 4 weeks, total) even when the external supplier's on-time performance is 80 percent.

## Multiple Operations with Long Processing Time

In a few unusual situations, there may be more than one long operation, but all the work centers may have a large amount of excess capacity. In such cases, the total processing time will be long compared with the standard lead time. Maintaining relatively high excess capacity is necessary for ensuring due-date reliability. In projects, this is normal (even though the people involved strive to hide their excess capacity). To apply S-DBR in this situation, several requirements must be met:

1. The routing of every order must be a straight sequence of operations without assembly of several components (which are also produced in the shop).
2. The excess capacity on all resources must be large—50 percent or more, especially at the resources with the long-time operations. This ensures relatively small queues (provided proper regulation of material release).
3. No multitasking is permitted.

Multitasking is the practice of working for some period of time on a task, but before completing it, that task is set aside and another task is begun. The new task typically comes from some other project. The resource then works on that task for some time and will then either switch back to the original task or go on to another—again, without finishing. In other words, the resource is faced with multiple "open" tasks simultaneously. Multitasking is essentially the opposite of having a sequential queue of work. The resource tries to do everything at the same time, with the net results that each task takes far longer to complete than it should. In manufacturing, a multitasking "culture" doesn't exist because the requirement for setup causes operators to work continuously on one operation until completion, and only then set up for another order. However, some manufacturing situations start to look similar to projects. In most of the cases that lie between pure manufacturing and pure multiprojects, the work culture is still based on manufacturing, meaning no multitasking. In some such mixed situations where a team of workers is assigned to do one specific operation, it could be that the team could be split in order to advance two or more orders. Such a splitting is essentially multitasking.

When we adhere to the three conditions above, we can maintain a time buffer of only three or four times the total processing time. Effective buffer management makes it work. The prevalence of spare capacity makes queues of more than two orders is rare for any resource. Total waiting time is commensurately limited. S-DBR implementation in this case is straightforward and should work successfully.

Note: Don't assume that the buffer should be only three to four times the net processing times. When there isn't much excess capacity, waiting time becomes much more significant. Don't try to create a formula to size the time buffer. The data needed to establish the real-time impact of excess capacity on a manufacturing floor are not usually available. As a rule-of-thumb, reducing the current lead time by half is, in most cases, effective enough to begin with.

## Ultra-Large Orders

Ultra-large orders are a pain in the posterior. Such orders could require more than 20 percent of the CCR's monthly capacity and of other resources as well, while an average order takes less than 1 percent of the monthly capacity of the CCR. They could create temporary bottlenecks at resources that are usually non-CCRs. They also cause other orders to wait excessive amounts of time for processing. Ultra-large orders also hit raw materials very hard.

Why are such orders received in the first place? This is a very important question, because before we establish rules to deal with the problem, we should determine whether we can prevent the problem from occurring. Here are some possible reasons for ultra-large orders:

1. The client's purchasing manager decides to optimize the purchasing work load by issuing fewer, but much larger, orders covering several months of need. Price reductions for larger volumes make ordering very large quantities even more desirable for the client.
2. A client's lack of confidence in the supplier often translates into larger orders submitted well before the client really needs it.*
3. A client really needs the full order. It could even be that for the client the order is not very large. Maybe what we produce are raw materials for our

---

* "All other things being equal, I'd rather have them at *my* house than your house"!

client, and the order is for a quantity the client requires for its next batch. This situation is not very common, or we would see such orders frequently. Nonetheless, it could happen if our client produces a slow-moving product and the client's production manager decides to do a large batch so they won't have to set up for and produce it again for a while. In other words, the impact of the client's obsession with "efficiency" falls on us.

4. The client is a distributor who requires stock at a new location.

A quick examination of these justifications is that the fourth expresses a real need. The third might not be a real need, but to change that perception is beyond what we could expect from the salespersons. The first two reasons reflect client perceptions that can (and should) be addressed in ways that don't require ultra-large orders. These ways require convincing the client to break the full order into several deliveries. We intentionally use the word "deliveries" here. We don't necessarily ask the client to subdivide the order itself. Thus, the client can enjoy the (perceived) advantages of one large order without imposing unnecessary pressure on the manufacturer's operations. Fulfilling an order in several smaller deliveries makes this possible.

Therefore, rather than routinely accepting ultra-large orders for a single delivery, first try negotiating with the client to split either the order or the deliveries. Even if the client doesn't agree, split the order internally yourself. Set some dummy due dates for the smaller-sized orders, and make the last one the due date committed to the client. If you don't do this, and the whole order is released to the floor, expect some real headaches in production.

## Dependent Setups in S-DBR

Simplified Drum–Buffer–Rope operates on the underlying assumption that the actual sequence of work on the CCR (or other resources, for that matter) doesn't adversely affect capacity in any significant way. As long as this assumption remains valid, following the priorities determined by market demand is sufficient for our scheduling. Using the planned load to establish safe dates for promising delivery of new orders, and for monitoring the load on the CCR, is sufficient for high production floor reliability.

Let's demonstrate the partial reliance of S-DBR on being able to change the sequencing of work without waste of capacity. Suppose the CCR has to process five different orders: Order1, Order2, Order3, Order4, and Order5. The due dates are such that Order1 is due the earliest, then Order2, and the rest. Suppose now that for whatever reason, such as a quality issue that required rework by an especially skilled person, Order1 is late and not yet at the CCR site. However, Order2 and Order3 and Order4 are already at the CCR site. What can be done? The most obvious decision is to process Order2. Subsequently, if Order1 shows up before the CCR completes Order2, then Order1 would be next, otherwise the CCR would proceed to Order3 and so on. But what if by changing the sequence from Order1 first to Order2, we cause a significant waste of the CCR capacity? If that is the situation, then we should have made sure that Order1 arrived on time. In other words, *we should have protected the CCR's sequence.* This is what traditional DBR does: carefully planning the sequence of work done by the CCR and protecting it with the special CCR buffer. If failing to follow a certain sequence at the CCR causes wasted capacity, it is usually an indication of sequence-dependent setups. Let's address that particular situation in detail and specify when it actually applies.

In a sequence-dependent setup, the time to complete setup depends on the nature of the *previous setup* at the same resource. Paint manufacturing is the classical example. When an order for darker-colored paint follows an order of lighter-colored paint, setup time might not be affected much. But white paint is more sensitive to contamination. If an order for white follows one for black paint, machine clean-up between color changes may be considerably longer.

Sequence-dependent setups are common in the chemical industry as well. Tanks often must be cleaned between different product batches. How thorough a cleaning is required depends on both the previous tank contents and the new contents to follow. Dependent setups may also pertain when changes in the size or width between successive items require sensitive tuning. Small changes are likely to require much less tuning than large ones. Many dedicated production lines, especially automatic mass production lines, are subject to dependent setups.

When dependent setups happen, it leads production planning to a "preferred sequence"—one that minimizes the total number of setups. Let's consider a simple example.

Suppose we have to produce four different products—P1, P2, P3, and P4. Table 5.1 shows how setup time (in hours) varies from the preceding item (the row) to the next item (the column). Thus, setting from P2 to P1 takes three hours, while setting from P1 to P2 takes only one hour.

When saving setups is important, the preferred sequence should be: P1-to-P2-to-P3-to-P4-to-P1. If we start with P1, we must complete the orders of P2 next, then P3 and P4, and only then go back to P1. But such a preferred sequence might

**Table 5.1  Dependent Setups (In Hours)**

| From \ To | P1 | P2 | P3 | P4 |
|---|---|---|---|---|
| P1 |  | 1:00 | 2:00 | 3:00 |
| P2 | 3:00 |  | 1:00 | 2:00 |
| P3 | 3:00 | 2:00 |  | 1:00 |
| P4 | 3:00 | 3:00 | 2:00 |  |

not be aligned with S-DBR/buffer management priorities. Suppose we have an urgent order for P1, then an order for P3, then another order again for P1 (but less urgent than the preceding two orders). The preferred-sequence logic discourages breaking the sequence just because of an urgent requirement for an out-of-sequence order.

Now if the manufacturing process has significant excess capacity, or if the variability of the setups is not large, then S-DBR is still viable because the dependent setup resource won't run short of capacity, even if several additional setups are done.

The logic of S-DBR requires that the processing sequence accommodate the delivery sequence demanded by the market. If we follow a preferred sequence to minimize the setup time, response to market demands will be much slower. We must consider the time required to cycle through all four products, say, for example, the time after leaving P1 until P1 can be produced again. This calculation is crucial in promising reliable delivery.

Using the above example, suppose that an order for P1 has been received when the CCR is processing P2. This means the new P1 order will wait until the CCR processes all intervening orders for P2, P3, and P4—whether they are as urgent as the new P1 order or not. Suppose this takes six weeks. (The term *cycle time* refers to the average time it takes from beginning production of an item until that item begins production again.) Now if another order, this one for P2, shows up while the CCR is set up to do P2, it might be delivered very quickly, perhaps less than a week. But this is purely a matter of lucky timing because the unfortunate customer for P1 has to wait for six weeks. Consequently, customers should know that when they place an order, it could take as long as six weeks for them to receive it. Of course, if the whole cycle time of going from P1 and back to P1 is only one week, then a safe due date of two weeks, or even less, could be attractive enough for the clients and S-DBR can be implemented without any problem.

When a required (or desired) sequence of jobs consumes too much capacity, then production must follow the preferred sequence. This could create problems for the S-DBR methodology because of two different obstacles:

1. The safe time the company should quote to reliably fulfill its promises might be somewhat longer than the cycle time. Maybe this will be acceptable to the customer. If so, it's not a problem. But it unquestionably makes the company vulnerable to competitors who can be more responsive. For example, presuming that all competitors are subject to the same kind of dependent setup situation, one who decides to limit its offerings to a lesser number of items would be able to realize a much shorter cycle time.

2. There is no practical way to expedite. It might be possible to deviate from a desired production sequence to do, just once, a longer setup (in other words, to complete a larger, combined batch). But doing so could adversely affect promised delivery dates for other orders. What if, while producing P3, there was a pressing need to switch back to P2? If the preferred sequence was

resumed at that point, it could be significantly more than the usual six addi-tional weeks until the CCR returns to producing P1. Clearly, in a dependent setup environment, we can't rely on the ability to expedite when necessary.

Dealing with urgent orders is the chief problem in dependent setup situations. If the CCR is forced to go through a predefined sequence—the preferred sequence—then the exact current state of the CCR within that sequence affects the earliest delivery time we can quote. This is the essence of the proposed solution.

A sequence-dependent setup seriously limits production's ability to respond quickly to changes in demand. Any competitor that could either develop a tech-nological solution or less expensively add another unit of the resource would have excess capacity and more flexibility to maneuver within the preferred sequence and could gain a decisive competitive edge. Consequently, as with the paint manu-facturing process mentioned above, it's worth exploring whether an engineering change, a technology change, or a process redesign might eliminate the depen-dent setup. Removing the limitations entirely is obviously better for flexibility and competitive advantage. However, if this isn't possible, the following solution is a second-best alternative.

## Sequence-Dependent Setups: A Proposed Procedure

When we must live with a sequence-dependent setup, it's beneficial to have the best possible procedure for managing the situation, given the circumstances. Due to the additional complexity of the situation, the proposed procedure is not as simple as S-DBR. Any reader who does not have special interest in the particular situation of sequence-dependence setups can skip this part.

The most significant impact on the length of the dependent setup is usu-ally caused by a certain product parameter, such as color, a certain chemical substance, size, or width. Let's assume that the key difference between products D1 and E1 is the width. Now what happens when D1 is not a specific product, but a *category* of products, all of the same width, but which have some other characteristics that make them different? There could be some setup involved when processing product D11 and product D12, but that setup time would be relatively short and not bound by the sequence, as long as they are products of the same width. Then we have E1 as part of another category of products with a different width.

In most cases, the major impact on setup time occurs between categories of products. Within a category, the setup time is relatively short and dependency on the sequence may be less critical. This characteristic will likely define the preferred sequence to be only between *categories* of products. The cycle time then becomes the time between producing a certain category and the next time the line (or a machine) will be set up to produce that particular category. Within a given category, there would still be enough flexibility to manage changes in product sequence.

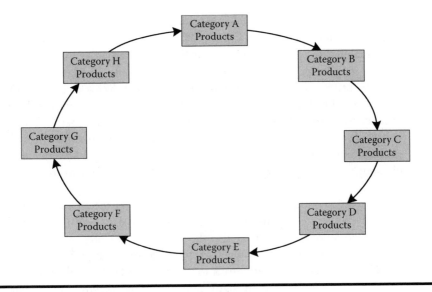

**Figure 5.10   Sequence-Dependent setups: A full-cycle wheel.**

We can depict the sequence-dependent setup situation as a wheel that establishes the preferred sequence between the categories (Figure 5.10).

So for example, if a certain order for F19 has penetrated the red buffer zone, but the rest of the F-category products are not in the red, then once the production line is set to the F width, F19 would be the first product to be produced.

In Figure 5.10, we have eight categories of products. In this case, the preferred sequence happens to be a clockwise rotation through the wheel. It's also possible to have a sequence within a category, but because deviation from it might be less critical, it might not be necessary to include it in production planning.

A dependent setup situation drives planners to accumulate all required orders belonging to a given category together, so they can be produced during the same time period within the cycle. But there are ramifications in doing this. Some of the immediate ones are:

■ For new orders, safe-to-promise delivery dates for some might be in the very near future while the safe-to-promise date for others might be much farther out. If, for example, we're now already set up to produce Category A, we know that orders for products in Category B will be produced relatively soon, but orders for Category H will be significantly delayed.

■ The farther into the future the safe-to-promise date is, the less certain we are of achieving it. There is a higher probability that normal general-cause and special-cause variation in the production process for Categories B through G could delay delivery of Category H products well beyond the original safe-to-promise date. Also, additional new orders for the B to G category probably would be added to the production thus delaying further starting to produce category H products.

■ The capability to expedite is very limited. Suppose an order for Category F (in the preceding cycle) was inadvertently overlooked and has now become very urgent, and we are now working on Category A. What can we do? One option is to break our current setup, shift immediately to Category F, and produce that order. This, of course, imposes an extra setup that violates the preferred sequence. Much more setup time would be required than we would experience if we respect the sequence of the wheel. But if we have the luxury of enough excess capacity to offset this waste of capacity, what should we do now?

  – We could continue in Category F, so as not to waste any more capacity. However, if we do this, we will cause a huge delay in all the orders for Categories of B through E (also the remaining of Category A that were not produced before the urgent setup change to Category F).
  – We could change the setup back to Categories A or B, but then the waste in capacity is much more significant. We could certainly do it, but it would be best to contemplate it only in extreme cases.

Whether we operate primarily in a make-to-availability (MTA) environment (which we discuss in Chapter 6), make-to-order (MTO), or a hybrid of the two, we must fix the length of the cycle—or at least determine a reasonable maximum cycle length. And even this isn't enough. We must determine the time allowable for each category as well. Otherwise it won't be possible to offer safe-to-promise dates or to maintain an acceptable replenishment time for stocked items. Clear guidelines are required for whether orders can be added to a particular category within the current cycle, or whether they must wait to be included in the next cycle.

To manage dependent setup production, we recommend the following procedures:

1. *Fix the maximum time for a cycle.* To determine what this time should be, consider how to balance between the overall capacity requirements of the CCR—including the setup times along between components of the category wheel—and the need to react to the market as fast as possible and make a decision. The decision fixes the cycle time that is not too long for the market and not too wasteful of capacity because of the heavy setups.
2. *Establish a buffer for the cycle.* Include within the maximum cycle time 15 to 20 percent of the total time as a buffer. For instance, if the cycle time is fixed at 15 days, then plan only for up to 12 days, leaving another 3 days for truly urgent orders. This buffer covers for fluctuations in product demand throughout the cycle. Let's call this the *cycle buffer.*
3. *Apportion the cycle time.* At the start of a production cycle (sometimes called a *campaign*) we have about 80 percent of the full cycle time available to apportion after establishing the cycle buffer. That 80 percent of cycle time should

be assigned proportionally to the categories based on the average time needed for each category. Of course such assignment is based on average demand observed in the past, but we also leave enough flexibility to deal with extra demand that may happen for some categories and less-than-average demand for others. This should produce the *assumed* latest finish time for each category. The start of the next cycle is the assumed end of the cycle plus the cycle buffer. Thus, category A might get two days of the cycle, then category B only one day, while the next one, category C, gets three days (from day 4 of the cycle until the end of day 6).

4. *Compile all firm orders within their designated categories.* Then calculate planned load for each category, including all stocked items. For instance, Category B, which has only 24 hours (one day), starts with a setup of 6 hours, and then all the existing orders for B products join the planned load for Category B.

5. *Insert additional orders as capacity allows.* As long as planned load is less than the portion of total cycle time allotted for the category, most additional orders can be inserted in a particular category "worry free." But once the planned load for a particular category equals or exceeds the category's allotted time, the cycle buffer is penetrated (for example, when additional orders for Category B brings the planned load for B to 30 hours). The cycle buffer exists to permit adding *some* orders to a category that might have more demand than can be accommodated in the average time allotted for it. When no orders penetrate the red zone of the cycle buffer, adding urgent orders makes good sense. Suppose the cycle buffer is three days and right now the overflow of orders beyond the average allocation time is 36 hours—one and a half days—which means the cycle buffer is in the yellow (50 percent of the buffer being consumed). But when the cycle time is in red—if, for example, 60 hours have been consumed out of the cycle buffer—more careful consideration is required. We want to avoid causing the entire cycle to exceed the planned maximum duration. Otherwise, we place the next buffer cycle at risk. If the cycle time takes more than the total time, including the cycle buffer, the next cycle will start late. The committed due dates for the orders to be filled by next-cycle-production would be compromised. In summary, when more orders arrive and the cycle buffer is already in the red, then most new orders should be scheduled for the next cycle.

6. *Recalculate the cycle start and end times.* When production in a category actually starts, the schedule for the rest of the cycle is recalculated. A category that finishes short of its allotted time adds to the cycle buffer, and the whole subsequent schedule is commenced earlier by the same amount.

7. *Calculate a safe-to-promise date.* Begin with the scheduled end time of the category, then add the cycle buffer plus one-half of the production buffer. The production buffer is the normal average duration used from material release until completion.

8. *Determine the scheduled material release time.* The scheduled release time for any given order is the most current *assumed start time for the product category* minus half of the production buffer.

   a. Even when the CCR is the first resource, we recommend having a defined time to ensure the materials are ready for release.

   b. The planned load is only a rough estimate of available capacity. It can't serve as a category schedule because there might be a lower-level preferred sequence within a category. Consequently, the trigger for material release is the time that setup begins for the category as a whole.

## Summarizing the Complications in Using S-DBR

Sequence-dependent setups represent the one environment where the S-DBR must be substantially changed, making it closer to the traditional DBR by having to adhere to a predetermined schedule of the CCR. But the complications described in this chapter demonstrate that the simple concept is still valid: *it is enough to manage a MTO environment based on the most straightforward pull concept of the flow:*

■ Every order is protected by a time buffer, usually just one, but in some cases two connected to each other.

■ All resources, including the CCR, should be subordinated to the market by following a general priority based on how much time has been consumed so far relative to the time buffer.

■ Red orders require intervention and managerial efforts. Managers are advised not to intervene with orders that are not already in the red.

■ Planned load is the main tool for monitoring capacity, and it dictates the safe dates that are the earliest delivery dates that might be offered to the client.

■ The close collaboration between Sales and Operations is mandatory for achieving the reliability and agility that could be used as a significant competitive edge.

The complications to S-DBR described in this chapter primarily affect make-to-order situations. In the next chapter, we examine the application of S-DBR in a make-to-stock environment.

## Reference

Rinehart, James, Christopher Huxley, and David Robertson, *Just Another Car Factory? Lean Production and Its Discontents.* Ithaca, NY: ILR Press, 1997.

# Chapter 6

# Using Simplified S-DBR for Making to Availability

## Contents

The focus of Theory of Constraints (TOC) industrial applications—its bread and butter—is on the reduction of inventory and facilitating its timely flow through the shop as the means to increase throughput.* However, the lessons learned from successfully applying TOC tools go well beyond inventory reduction and improved flow. The ability to recognize areas of opportunity or extension fits snugly into the mainstream robust problem-solving mission that TOC practitioners have come to accept. It's reasonable to assume that some capability must exist beyond make-to-order (MTO) applications in manufacturing. In this chapter, we will identify some of the specific needs that have led us to conclude that making to availability (MTA)—a special case of making to stock (MTS)—may, in fact, supplement the benefits already realized by firms with MTO applications. In addition, we will offer some warnings and suggestions for implementation.

Simplified Drum–Buffer–Rope (S-DBR), like its traditional predecessor, was originally intended to help the companies using it shift from producing for finished inventory to filling firm customer orders. The underlying assumption behind this intention was the idea that, in many situations, the faster production flows possible with DBR make producing to order feasible and allow significant reductions in finished goods inventories.

This is still the best use of both traditional DBR and S-DBR. Small to medium-sized companies are usually able to realize the best benefits for making to order with DBR. Making to order has been less practical for larger companies, which often sell their products through networks of retailers, many of which are served by regional distribution centers. For such companies, everything they produce is always filling stock somewhere. And such manufacturers are inevitably just one link—albeit a very important one—in a fairly long and complex supply chain. Thus, the need to consider how to use S-DBR in a make-to-stock environment was a natural evolution in production thinking. In this chapter, we will explain how S-DBR can be used to make such supply chains robust; in other words, how to also use it for making to stock.

---

* Remember, in TOC "throughput" is not the same as "output" (volume of units); it represents the financial value of that output.

# Why Make-to-Stock and How It Differs from Making to Order

Making to stock means that production is begun without a specific customer order. This definition also includes producing intermediate parts or components to be used once we have customer orders in hand. It also includes producing to a forecasted demand, or producing according to a min–max algorithm.

In the best of all possible worlds, we would always prefer manufacturing to fill a firm order. After all, what's better than a client with a check in hand before any purchased material is released to the shop floor?

Yet making to stock is a tempting option for some manufacturing companies, especially those that sell some standard products. In some cases, some of the products might be standard enough that stock can be produced even without a customer order, because customer demand leading to a sale is highly probable sometime in the not-too-distant future. However, in most organizations the make-to-stock option typically results in the production of too much stock.

Most service organizations need a customer to generate a service, thus they accept the need to maintain a significant amount of excess capacity. But in manufacturing, the common thinking is that making to stock can smooth the production load throughout the year. Consequently, whenever there are no urgent customer orders, companies are reluctant to see "valuable resource capacity" wasted, so they produce products they hope (or expect) to sell in the future.

There's a real dilemma here. This desire to smooth the load and ensure high resource utilization runs up against the problem of producing something that is definitely not needed right now, and may not be needed in the future (Figure 6.1).

There are two problems associated with producing what is not really required at the moment. First, it represents an investment of money—purchasing materials—that might not return sufficient value if the finished goods are drastically marked down or not sold at all. Second, if you have a capacity-constrained resource (CCR), by producing to stock you waste capacity that could be better employed making other products that might currently be in demand by customers. This often happens when production is ignorant of exactly what product is experiencing urgent demand and which isn't, or for which demand is not as urgent. If a CCR isn't currently active, you might actually create one by producing to stock, filling the shop with work in process (WIP), and creating confusion about priorities. Creating a CCR unnecessarily may eventually cause long lead times and bad due-date performance—not particularly good for maintaining favorable market demand for the stock produced.

How serious is this conflict? The fact is that most of the time the benefit from smoothing the load is minimal and not really required. If enough protective and excess capacity exists, why would one squander the one asset—excess capacity—when one could, with proper use of that capacity, ensure meeting delivery commitments and provide the ability to respond much faster? While making to stock

**Assumptions:**
1. Without making to stock there is a lot of idle time.
2. What is made without a firm order can always be sold.
3. What is made without a firm order can always be sold *at full value.*
4. Making to stock is the only way to assure high resource utilization.

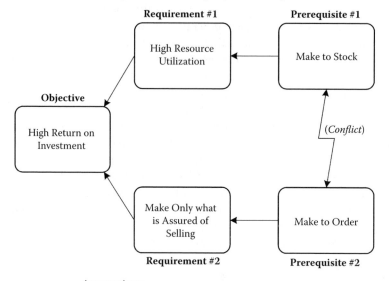

**Assumptions:**
5. Customer demand is uncertain.
6. Significant finished inventory remains unsold.
7. Unsold inventory is either deeply marked down, sold at a loss, or scrapped.
8. Making to order is the only way to avoid inventory obsolescence.

**Figure 6.1    The make-to-stock dilemma.**

is common, it's a counterproductive practice stemming from the logically flawed notion that an idle resource constitutes a major waste.

## Exceptions to Making to Order

One instance in which smoothing the load by producing to stock might have merit is in preparing for a peak demand period. In peak demand periods, a CCR might quickly become a real bottleneck. The only other alternative for a make-to-order environment is to commit to much longer delivery times when the peak demand begins to overrun the planned load. In other words, promise longer delivery delays, and actually deliver that way. We will return to the case of seasonality and the situation in which a bottleneck emerges during a peak load.

However, there's another situation in which make-to-stock is really required and beneficial: *when the tolerance time of the clients is shorter than the production lead time.* In this case, producing to stock is absolutely necessary or sales will be lost. This often happens when some competitors in the market have already seen an opportunity to increase their market share by offering much faster response, perhaps offering guaranteed availability of products. In other words, the competition wins a big part of the market by deciding to make to stock.

The key point is that it *is* possible to gain a competitive edge by making to stock. It enables businesses to offer complete, almost instantaneous availability of all products to clients. But the competitive edge accrues *only when the client realizes real value from such full availability.* This is certainly not always the case. Having a supplier that can provide immediate availability is convenient and desirable, but it doesn't add real value unless the client knows ahead of time that promising immediate availability can be used either to increase future sales or to charge more for current sales.

Even this capability may not be enough, because clients who need that immediate availability usually manage these items as their own stock. Most manufacturing companies do this. They don't rely on fast delivery from the supplier, nor do they expect their client to wait long enough to absorb both supply time and production time. Consequently, most companies manage their own materials as on-hand stock. They may see some benefit in our offer of immediate supply, but would it be enough benefit to justify converting from low-risk, make-to-order to high-risk, make-to-stock? Would this capability really improve sales or justify a significant price increase?

## Who Stands to Gain?

Which clients stand to benefit from immediate supply on demand? The most obvious ones looking for immediate gratification are end users (consumers). Producers of consumer products, especially those already producing to stock, also have a lot to gain if their supply of raw materials for production is always assured and immediate.

We have actually touched here on the reality of any supply chain, but let's concentrate on those that sell consumer goods. Consumers don't want to wait for the items they buy, and they want them to be in close proximity. The mission of distribution companies is to bring the items to be sold to retailers, who are close to the potential consumers. Distribution chains buy such items from the producers of finished products, and most of the time they do so by issuing orders to producers. So in effect, producers actually do produce to order in these situations.

But is this really the case? On one hand producers receive orders, yet on the other hand future sales depend on how clients manage the stock provided by the producer. When clients—distributors, wholesalers or retailers—don't order enough (possibly because they underestimated demand), they experience shortages. If this occurs, *both the client and the producer lose sales.* When clients order too much, their stock moves slowly, so they delay their next order, possibly for a long time. And maybe they order substantially less next time. Obviously, this doesn't benefit the producer.

So producers, distributors, wholesalers, and retailers are really all in the same boat. They all depend on a sale to the end user. When such sales don't happen, for whatever reason—lack of availability or inferior competitive position—the adverse impact eventually ripples back through the supply chain. But even if consumer demand for products doesn't lag, mistakes by one or more link in the supply chain can degrade sales.

In recent years, vendor-managed inventories (VMI) have become popular in some sectors. Producers of consumer products offer to manage clients' stocks for them. Theoretically, this allows clients to concentrate on selling their goods as fast as possible, while the producer ensures that there are always products available to sell. VMI is especially effective for maintaining the variety of components for large assembly plants. In such an environment, producers of the components take the responsibility for the stock of their components at the site of the assembly plant. This approach has the potential to provide producers a competitive edge and also to enhance sales. If producers are as agile as required, they should be able to ensure that no shortage of their products occurs. But doing so requires a deft balancing act between having too much stock in the pipeline, causing a whiplash effect, or too little, resulting in lost sales opportunities.

Some very large companies, such as the largest car producers, computer manufacturers, and certain very large distributors, have forced their suppliers into VMI. This kind of heavy-handedness is not conducive to a good "win–win" relationship. But it does illustrate two important facts:

- Certain clients benefit significantly by having assured availability of their supplies.
- Certain companies are forced by their clients to make to stock for them.

Both testify to the reality of situations where making to stock is required. However, even when it isn't forced, making to stock warrants consideration. Naturally, when considering a shift from make-to-order to make-to-stock, it's important to carefully consider all the operational ramifications.

## Making to Availability

Making to stock is a well known and widely accepted concept in manufacturing. However, within the general concept of making to stock, we introduce a new subset: *making to availability* (MTA). MTA is a manufacturer's general declaration to provide immediate supply whenever needed. In certain cases, such a service declaration might specify certain customers and specific sites where the availability would be provided. But just as often, MTA is a broader commitment to the market at large. As we see it, MTA is a sensible, justifiable reason for making to stock. In many cases, though, ensuring availability isn't part of making to stock. Consequently, it's important to differentiate between the two (see Table 6.1).

We will examine the potential impact of TOC on global supply chains in more detail later in this book. For the time being, however, we must establish the case for making to availability.

## Differences between Making to Stock, Making to Availability, and Making to Order

Let's suppose that you run a manufacturing company that has historically produced to order, but recently your customers have been pressuring you for faster response. Because some of your products are largely standard (meaning the same item can be sold to many potential customers), you feel compelled to consider producing to stock, even guaranteeing availability of some of your products. What changes must be introduced to your operations to do this?

It seems a bit incongruous that Manufacturing Resource Planning (MRP) and other popular production planning methodologies treat making to stock practically the same as making to order. Once a production order has been planned, it's impossible to tell whether it satisfies a specific customer need or is intended for stock. All production orders specify only a date for completion. But isn't there something strange here?

### Due Dates

When an order is intended for a specific client, we expect to see a specified time for completion. And sometimes the customer's delivery due date differs from the production completion date because time for transportation to the client must be included. But why should an order intended for stock carry a due date? Who can possibly determine when some customer would actually require the item?

It's apparent that almost all manufacturing companies assume that every production order must have a completion date. For years, everyone has accepted the way work orders are generated in MRP: Different customer orders for the same item (perhaps a common part required by different end products) are merged into a single production order, despite the fact that due dates could be quite different. An emphasis on local efficiency drives this behavior—the idea that setup time must be reduced, regardless of whether the time saved delivers any real value.

But when we produce a work order combining multiple customer orders having different due dates and a work order for some units intended for stock, the idea of single due date for the work order has actually lost its meaning as far as delivering good service is concerned. The only priority seems to be exerting pressure on the manufacturing floor to perform well, where "performing well" means meeting due dates and efficiency objectives. Moreover, due dates for stock orders are usually

**Table 6.1 Types of Manufacturing Strategies**

| Production Policy | Customer Relationship | Characteristics |
|---|---|---|
| *Make-to-Order*<br><br>A policy to produce only when a customer order is received. Thus, both the required quantity and the delivery date are fixed. | ■ Ad hoc | 1. One-time or multiple times<br>2. For discrete customer need (sale guaranteed)<br>3. On demand (no regular schedule); production commenced only on receipt of order<br>4. Shipped immediately upon completion<br>5. Defined quantity<br>6. Defined delivery date<br>7. No guarantee of subsequent orders<br>8. No manufacturer inventory |
| | ■ Long-term contract | 1. Multiple times<br>2. Regular schedule (interval)<br>3. Shipped on scheduled date<br>4. Exact quantity usually defined per shipment<br>5. Defined delivery date<br>6. Guaranteed (negotiated) price<br>7. Guaranteed subsequent orders (for life of contract) |
| *Make-to-Stock*<br><br>A policy to produce without a firm customer order, i.e., to finished inventory. No fixed quantity or delivery date. | ■ To inventory | 1. Recurring<br>2. Production initiated without a definite customer (sale not guaranteed)<br>3. Fulfills a producer's need and decision |
| | ■ Make to Availability | 1. Fulfills a commitment for continuous near-immediate availability of specified SKUs<br>2. Promised to a specific customer (or to all customers)<br>3. Customer orders not filled until submitted; regularly scheduled or on-demand<br>4. Stored at producer's site (producer owns inventory) |
| | ■ Vendor-Managed Inventory | 1. Same as Make-to-Availability (#1 – #3)<br>2. May be stored at producer's site or customer's site (customer's owns inventory) |

determined by the standard production times, so the relevance of due dates for a make-to-stock order is very low indeed.

On the other hand, if a production order has no due date, how can we be sure it will flow to completion and not be rat-holed or shunted aside along the way? Such an order wouldn't have any priority, thus, in any conflict with an order that *does* have a due date, the order with the due date would seem to have a higher priority. Might this always be the case? We will revisit this question when we address buffer management in making to stock.

## Determining Priority

In producing to stock, the point is that, while we must find a way to determine the *quantity* we want (material release is determined by the quantity of finished units required at completion), there is really no pressing need to determine the *completion time*. The relative urgency of the order might even change while the order is in production. Could the assignment, or an update of priority for some orders, result in other orders being stuck somewhere on the shop floor? Perhaps, under two conditions:

1. A low priority given to an order may not allow the order to progress at the rate needed. The updated priority is kept very low and most other orders have higher priority, thus that particular order is stuck.
2. A capacity-constrained resource (CCR) is active on the production floor. All other resources should have idle time periodically, meaning that a low-priority order should, at times, be the only order at a nonconstraint. In this case, that resource should process it.

If both conditions are present, the order really is of low priority. If a CCR is loaded with higher-priority work, it's appropriate to let that order wait, perhaps even for a very long time. This would also be an indication that sales are very low, perhaps nonexistent, and there would be no real need for that order. So one critical ramification of making to stock in S-DBR is that the production orders have no due dates; however, they *should* have a priority that is constantly updated.

## When to Initiate

Another critical ramification is that in making to order, especially in the S-DBR environment, the receipt of a customer order is the trigger that initiates a production order . But what should the trigger be in making to stock? In MRP the most common approach is to plan the rolling master production schedule (MPS)—the planning mechanism of MRP that generates the production orders—once in some designated time period, such as weekly, and to base it on a forecast. A rolling plan means that although the actual horizon of the master production schedule might easily be several months, the MPS is recalculated every week, especially as the forecast changes. What could possibly be wrong with such a sensible approach?

## Forecast-Based Planning: Unavoidable Consequences

Forecasting is a mathematical–statistical process intended to determine information about the future. Often overlooked is the fact that this is only *partial* information. A manufacturer making to stock must know the volume of future sales so the right products can be produced at the right time. Forecasting seems to provide the ultimate answer to the question: "How much to produce and when?" But a major problem with forecasts is that too many people ignore the "partial information" concept in forecasts. We can confidently say that the use of forecasting in most managerial areas is rife with ignorance, and it unavoidably leads to wrong decisions, highly undesirable outcomes, and mediocre financial performance.

Let's consider a simple example. Suppose the forecast for the sales of a certain new book next month is 1,000 copies. Does this mean that if the actual sales next month turn out to be only 913 copies the forecast was "wrong?"

Many would disagree about whether 913 (rather than 1,000) might be considered "good enough" or not. Reality has certainly taught most of us the lesson that forecasts are not "accurate." But then forecasts were never *intended* to be accurate, so how large a deviation from the forecast would *you* consider to be "acceptable?" Five percent? Twenty percent? More?

The inconvenient fact is that in some cases, a forecast of 1,000 might be considered a good one if reality actually turned out to be 913. However, in other cases the same deviation might be considered a total failure of the forecast, probably caused by a conceptual mistake. The point is that there are no deviation standards for what constitutes a "good" or "bad" forecast. It depends on the circumstances of each situation. Obviously, when real demand turns out to be exactly as forecast, we could say that the forecast was a good one—but, that rarely happens, except by luck*.

How does statistics deal with an uncertain phenomenon, such as future sales? General statistical tools typically require at least *two* parameters to describe uncertain behavior: the *average* and the *standard deviation* (Figure 6.2). The first determines a central value and the second describes a nominal deviation from that value. We won't go into the abstract technical details. Suffice it to say that the idea is to determine a value and the most likely deviation from it as the best available *partial* information for making sensible decisions. Another way to describe uncertain behavior is through the "confidence interval"—the interval that contains the possible actual results according to the level of confidence required. For instance, we might wish to have the interval where in 90 percent of the cases the actual result

---

* "Even a blind squirrel finds an acorn once in a while."

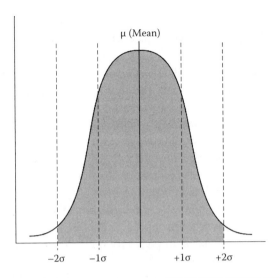

**Figure 6.2   Mean Forecast and Deviation.**

will be contained by the interval. Usually the confidence interval is determined by using the average, plus and minus some number of standard errors, depending on the level of confidence required.

In reality we might have only an approximate assessment of the average and an even less secure assessment of the standard deviation. We might still use the idea of the confidence interval to describe a reasonable interval of results.

Returning to our example, forecasting sales of 1,000 units usually means that sales are expected to *average* 1,000. But even if this forecast is accurate, it still doesn't mean that the actual result will be exactly the forecasted average because there is still variability present. On the other hand, if the forecast is *not* the outcome of a statistical model, but is based on the intuition of people, we may ask them to forecast a reasonable range of results.

Suppose that in our example the forecast is 1,000 books with an estimated standard deviation of 20 books. This means that we have about a 95-percent chance that the actual sales will be somewhere in the range of 960 to 1,040˙. The actual results—only 913 books sold—indicate that the forecast is significantly flawed. Sales that deviate from the mean by more than 4 percent are not very common. In other words, this is a very stable environment in which sales can be reliably predicted.

An intuitive forecast would say it is not reasonable to assume sales of less than 500 books, and also not reasonable to hope for more than 1,500 books sold. This may be a

---

˙ We have assumed here a normal distribution of the sales, which is not always the case. But it is still reasonable to expect that sales will range between the average and plus-or-minus two standard deviations.

more realistic forecast for one specific book in its first week of sales. Now we are more ready to base a decision on how many copies should be printed for the first week. The standard for accuracy of the forecast is reduced because we have increased the standard deviation to approximately 250 books for our interval estimate (Figure 6.3).

When it comes to such a decision, the average forecast is *not* really important. Printing exactly 1,000 books would be, in most cases, the wrong decision. There is very little chance that the customer base would request precisely 1,000 books. So the real question is which outcome would yield the least damage: *printing too many books or too few?* In the former case, after the sale, we would end up with unsold inventory; in the latter, we would lose an opportunity to realize more sales. Let's call overproduction a "type A" error, and underproduction "type B."

In most situations, one type of error turns out to be grossly more damaging than the other, so our decision should protect us from the greater damage. Because our printing is primarily in preparation for the first week sales, if actual customer demand turns out to be much lower than the projected average, we might be stuck with the remaining books for a long time until they are sold—if we can sell them at all. The damage is not just the cost of the paper and ink that were used in the printing, but that creating an excessive inventory of one book may actually harm the sale of other books. In short, it is difficult to assess the damage, but it is significant.

On the other hand, if demand turns out to be much higher than the forecast and the book is sold out, many potential customers might still want it after the big sale day, so printing more books after the first successful week would still

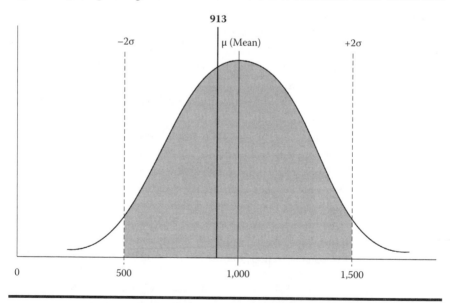

**Figure 6.3   Mean Forecast and Deviation (Example).**

capitalize on the first success, but maybe not to the same degree. We also face the difficulty of assessing the damage. But one might conclude that the damage from the type A error (having many unsold units remaining) is far greater than that of the type B error (being sold out, leaving some potential clients disappointed).

So how many copies should we print? Certainly not a thousand units. The size of the first printing should be targeted at the reasonable minimum units that we can expect to sell. To calculate the approximate number, we need either an estimate of the forecasting error or a "forecast" that doesn't provide *average result*, but the *minimum* that we could reasonably expect to sell (i.e., only approximately a 5 percent chance that actual sales would be less). Suppose this estimate turns out to be 600 copies. In this case the decision should be to produce 600 units, but definitely not 1,000.*

At the end of the day, we recognize that we can buffer ourselves from one side only. Thus, either we protect from shortages or we protect from surplus of inventory. Then, when it is clear to us what to protect, we do our best to establish the buffer at a conservative level. In TOC, we usually take for granted the protection from shortages, which is the right thing to do in most cases. Then most of the energy goes to preventing excessive inventories that do not add to protection and thus cause damage.

The disadvantages of planning production based on a forecast are:

- *Failing to ensure adequate protection from the damage of a wrong decision.* This in itself can create simultaneous shortages in some product lines and surpluses in others. It results from misunderstanding what producing to an average forecast value actually does.
- *Failing to account for the uncertainty in demand.* There is a cost associated with operating in an uncertain environment. Management should assess the possible costs and ensure that the organization can sustain them. When you rely on a single forecast value (average) alone, the potential cost of the actual deviation from the forecast is not even considered.
- *Failing to appreciate the importance of safety stock.* Even when safety stock— stock deliberately added in addition to the average consumption predicted by the forecast—is included in the plan, its validity is not monitored. Thus, whenever cash problems arise the safety stock is invariably reduced because no one can explain how necessary it really is.
- *Extending the forecast too far into the future.* Once the forecast is established, it's very tempting for management to apply it for longer periods. The motivation for doing so is the intent to make production "more efficient" through batching and by extensive stock buildup long before peak sales are predicted.

---

* In reality, it might not come to an either-or decision. Many retailers who underestimate the demand for a product on sale typically offer disappointed customers a rain check—a promise to deliver the product at the same sale price somewhat after the sale, when more can be produced against a somewhat more firm demand.

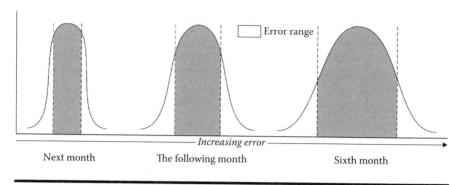

**Figure 6.4   Increases in Forecasting Error over Time.**

The misunderstanding of the role of the forecasting error blinds the decision maker from realizing that, the farther away they look, the higher are the forecasting fluctuations per month. In other words, the forecast for next month might fluctuate by 20 percent, but the forecast for month 6 is so far into the future that it could fluctuate by 100 percent. Thus, the planning flaws are getting very large (Figure 6.4).

■ *Turbulence caused by reforecasting.* The inaccuracy of longer-range forecasts eventually becomes apparent. The "band-aid" management typically puts on this problem is reforecasting—updating or changing the original forecast. New forecasts inevitably introduce changes because the new one always differs to some extent from the previous one. These changes can cause havoc in execution. Does this mean that reforecasting should be done more frequently to minimize the turbulence? Actually, frequent forecasting has a way of increasing instability. Remember W. Edwards Deming's admonition about "tampering" with a system that is, in reality, in statistical control (Deming, 1986). This phenomenon manifests itself in frequent setup changes in response to changes in the calculated average, despite the fact that it may still be within the control limits.*

The inevitable result of placing too much emphasis on forecasts is too much inventory of certain items, and insufficient inventory of others. This can be frustrating, especially for manufacturers who invest precious capacity of a constrained resource to produce items that may eventually remain unsold, or move very slowly, while they experience shortages of other items with significant demand. The inevitable result of reforecasting is even wider fluctuations and the sense of always being

---

* It must be emphasized that we are not arguing against setup changes, merely against setup changes that are undertaken for the wrong reasons (i.e., chasing forecasting changes).

in an emergency situation, trying to fix what appear to be the mistakes of the previous forecast.

What can a make-to-stock/make-to-availability manufacturer do? We know that the traditional use of forecasts leads to problems, but is there another way? It's certainly not practical to hope for a crystal ball.

## A Proposed Solution for Making to Availability

The "right" size of any inventory is determined primarily by two variables: demand and supply response time.

Demand includes both the average "pull" for an item and the range of fluctuation in that pull. Clients' perception of damage to themselves resulting from a shortage influences their demand. That perception is influenced by the client's potential to use alternative products or sources of supply. Another demand factor might be the clients' "tolerance time," and the fluctuation in that time. The tolerance time for most clients is likely to be larger than zero, so it might be possible to hold a little less inventory yet still provide the required service.

The supply response time is also referred to as the *replenishment time*. The definition of replenishment time is *the time from consumption of a unit of inventory until that unit is replaced*. The shorter the replenishment time, the less inventory clients must hold to guarantee their customers product availability. Here, too, fluctuations must be considered. But the "reliable" replenishment time is especially important. This is not necessarily the shortest time to replenish—it's the time duration the client can depend on for assured receipt of an item when it's really needed.

With demand and replenishment time as our entering arguments, we should be able to minimize the undesirable inventory conditions: too much inventory or too little (not necessarily the same items at the same location). Forecasts of demand are projections of an uncertain future. The time for replenishment occurs in the here-and-now, and affecting it is likely to be more within our capabilities than improving the accuracy of demand forecasts, or changing customers' service expectations.

## Basic Principles of Making to Availability

Five essential principles will guide our efforts to "bulletproof" availability by minimizing risk of damage from understock or overstock.

1. *Inventory and replenishment time are closely corrrelated*. Faster (shorter) replenishment times require significantly smaller inventories to ensure availability and avoid lost sales. Shorter replenishment times mean more accurate demand projections, too. Conversely, the longer it takes to replenish a finished stock, the greater the volume of finished stock required to maintain availability. The stock level required is roughly equivalent to the potential demand within

the time required to replenishment—in other words, consumption during the interval between the sale of an item and the time a replacement arrives. The stock in the system should be sufficient to accommodate the potential demand during that time. As replenishment times increase, more inventory is required to guarantee sales and ensure against changes in demand. Clearly, the implication is that if finished stock levels are to be minimized, production time must likewise be shortened as much as possible.

2. *Work-in-process supplements protection of availability.* While much of the direct protection of product availability for the market resides in finished inventory, production orders "on the way" are also part of the overall protection. To ensure availability, work-in-process is required, too, not just finished stock. The rationale behind this principle is that while work-in-process is not instantaneously available for customers, much of it is very close to being ready to go. Thus, the simplest effective way to ensure availability is to maintain some fixed amount of stock, both in finished goods and in the pipeline. Let's call this fixed amount of stock the *target level*. Though actual on-hand finished stock may fluctuate, based on variations in demand and production time, the system as a whole is stable. As finished inventory is consumed, completed production orders flow to finished inventory to prevent stock-outs, without accumulating too much finished inventory.

3. *Tomorrow will be similar to today.* This is the most basic of short-term forecasts. It's true not only for demand, but even more so for the combination of demand and supply. Consequently, unless we notice a very clear indication that there is a change in trend, we assume the currently established stock target in the system is the correct one to assure short-term availability. The immediate ramification of this principle is that each item produced to availability should have a target level of units that are either in finished inventory or in the production pipeline.

   Our reliance on work-in-process and day-to-day similarity (i.e., no sharp demand changes over one or two days) also means that whenever some consumption of finished inventory occurs, a new production order should be generated. We'll address this process in more detail later. However, the basic idea is to keep the replenishment time as short as possible, by responding to consumption very quickly, and by keeping the total stock in the system relatively constant (keeping the pipeline "pressurized"). There are other techniques for reducing replenishment time, but these are the essential manufacturing responsibilities.

4. *Status of finished inventory dictates production floor priorities.* Whether introducing new work orders into production or expediting work already in process, the size of deviations from target levels for each item determines which work orders should receive priority. To that end, we will define three basic priorities:

a. *Green.* If finished inventory is two-thirds or more of the targeted level, it will be considered to be higher than absolutely needed to support availability. This implies that replenishment needs are not urgent, but it also might indicate that we're carrying too much finished goods inventory. If replenishment can be achieved at a rate that exceeds normal demand, a high level of stock might not be required. If we should notice that there is work-in-process in production for an item for which finished inventory is "green," there is no urgency to this order whatsoever. This is not to say that it should be completely ignored, only that the priority of this particular order is low in relation to other orders.

b. *Yellow.* If on-hand finished inventory for a specific item is between one-third and two-thirds of the target level, we'll consider this normal, regardless of the fact that the term *yellow* typically connotes "caution." A yellow status implies that nothing is particularly urgent and that the inventory is not too high. The only message to production work centers is that the yellow status order has priority over green orders, but no more than that. There is no need for management intervention on yellow orders.

c. *Red.* When finished inventory is less than one-third of the target level, the risk of a stock-out is increased. Immediate action is required to restore the red inventory to yellow or green. Management intervention may be required.

5. *Stagnation is undesirable.* While we can expect different items to be in the red or green zones from time to time, "permanent residence" in either condition for too long is a signal that the inventory target level requires adjustment. If the item is red too much of the time, it means that we are consistently running too close to the ragged edge of stock-outs, and the inventory target level should be increased. If it's always green, it means that we are consistently holding more inventory than we need, and the target level should be decreased.

These five principles are the guidelines for an effective planning and control process to support MTA. While the first principle—the relationship between inventory and replenishment time—is somewhat independent of the others, it should serve as a reminder to continually search for ways to reduce inventory while still maintaining assured availability. From time to time, that first principle should be kept in mind when there's no easy way to provide fast production time, such as in the case of significant sequence-dependent setups. If you must live with a relatively slow response time to changes in demand, you may have to resign yourself to maintaining larger finished inventories and devoting constrained capacity to producing items that may not sell for a while. Understanding that relatively long replenishment time requires maintaining high finished inventory should motivate managers to find technological or other means to break through this limitation.

## Making to Availability: The Process

Let's lay out the procedures for operating a production process in a make-to-availability mode. These procedures will differ somewhat from those described for making to order in Schragenheim and Dettmer's *Manufacturing at Warp Speed* (St. Lucie Press, 2001).

### Step 1: Define the Initial Inventory Target Levels.

From our discussion of forecasting error, above, it should be clear that our objective is to ensure that we can meet almost any request for products while simultaneously avoiding excessive finished inventory. A natural outcome of the latter criterion is maximizing inventory turns. This means that we must find an appropriate level to set for finished inventory for every item we plan to offer with an availability guarantee.

The appropriate target level for each item maintained for availability should be equal to the highest demand we might reasonably expect to occur within a defined period of time. That period is defined by the average replenishment time plus a safety buffer to ensure against a reasonable delay in replenishment. This might sound a little complex, so let's illustrate with a numerical example (see Table 6.2).

We'll assume that Item-1 has an average demand of 60 units per week. The average replenishment time, equal to the production time, including the time it takes until the work order is released to production, is four weeks.

On average in a four-week period we would sell 240 units. But actually we must know how many units *might* be demanded in that four-week period and how long the replenishment could be. Let's suppose that occasionally the weekly sales could be as high as 100 units (400 units in four weeks). Should we set the target level for 400 units?

Even before we consider the possibility that replenishment time might exceed four weeks, we need to evaluate how likely it would be to experience four consecutive weeks of peak sales. If it's not likely, then what might we expect the four-weekly consumption to be? By either intuition or historical data review, let's assume that 320 units would be the four-week maximum.

Now, what is the joint probability of simultaneously experiencing such peak sales and longer than average expected replenishment time, meaning it takes longer than four weeks? If we have extensive historical data, we might be able to calculate an estimate, but as often as not we may have to use the TLAR method.* Let's say that, allowing for peak demand and an unexpected replenishment delay (a day and a quarter), we arrive at an estimate of 350 units.

Though this might seem complicated, a make-to-availability manufacturer must make this calculation for any item it produces to availability. The biggest challenge

---

* "That looks about right"

**Table 6.2  Initial Finished Inventory Calculation**

| Product | Maximum Weekly Demand | Average Replenishment Time | Estimated Maximum 4-week Consumption* | Additional Safety Factor for Delays in Replenishment | Initial Inventory Level |
|---------|----------------------|----------------------------|---------------------------------------|------------------------------------------------------|-------------------------|
| Item-1 | 100 units | 4 weeks (20 work days) | 320 | 1.1 | 350 units** |

\* *The maximum for 4 consecutive weeks has to be lower than four times the maximum sales.*

\*\* *We are not precise here. Technically, 10% of 320 is 32, which would total 352. We can round this number to 350, since the buffer is a gross assessment anyway.*

is deciding on a reasonable estimate of the upper limit of demand—maximum sales—within a defined time period. We must emphasize that the average maximum demand for 10 days should *not* be determined by multiplying the highest daily demand by 10. Doing so is likely to produce an unrealistically high figure.

There are two approaches that use a somewhat more simplified algorithm:

■ The maximum sales within *reliable* replenishment time
■ The maximum sales within *average* replenishment time *multiplied by an uncertainty factor.*

The first approach uses the term "reliable replenishment time." Reliable means we're highly confident that the resupplied items will arrive in that time. It's somewhat longer than average, but not the maximum. We're assuming here that we don't exaggerate the duration of the replenishment time. It shouldn't be the maximum time because it's unrealistic to include both the maximum demand *and* the persistence of that demand level throughout the maximum replenishment time. This approach best fits a situation where the total number of suppliers is limited, so there are not too many reliable replenishment times. It also assumes that the implementer can create a spreadsheet that considers the recent history of sales and determine the actual maximum sales within the estimated period of reliable replenishment time.

The second approach can be useful when there is no easy way to assess the reasonable high demand within a period, but a regular forecast that predicts average sales is available. The average of sales within the average replenishment time represents an estimate of overall customer demand from the time of one sale until its replenishment arrives. Now this quantity is not going to be high enough for our purposes. By definition, *average* means that a significant number of sales will exceed the average, so if we expect to protect our promise of availability, we would have to add a significant amount of safety.

However, the question is: *how much safety?* By what factor could we expect sales to exceed that average number? And if that happens, could the replenishment time also be slower than normal? All we have is the two general approaches. The important point, though, is to remember that there's no need to be precise. This is only a starting point. As time goes on, buffer management will indicate the need for changes to the target levels, based on the actual combination of demand and supply. We will discuss this issue in more detail later.

Therefore, let's not waste too much time defining initial target levels for inventory. Though we don't want to make a major mistake, starting with average demand within average replenishment time and multiplying it by a factor of 1.5 or 2 (or even 3) is a reasonable beginning. Our minimum should be a factor of no less than 1.5 when we are setting initial inventory for a central warehouse, which serves many distribution nodes. By accumulating at one point demand from various nodes, overall demand will fluctuate only moderately.* So multiplying the average by a factor of 1.5 is usually an adequate starting point. A client demanding daily replenishment might require a safety factor of 2 or 3, because variation is more pronounced on a smaller scale.

Returning to the example of having, on average, sales of 60 units a week and replenishment time of four weeks, then on average the needs are 240 units. Multiplying by a factor of 1.5, we arrive at a target level of 360 units as the initial stock, which is quite close to the 350-unit level reached by the more elaborate calculation. And who can really determine which buffer is more "correct"? Both seem close enough.

## Step 2: Generate the Production Order

Once the inventory target levels are established, production's job is to maintain them. This means that whenever the total inventory for any item (the total of finished stock plus open production orders) is below the target level, a new production order should be generated immediately.

That sounds simple and straightforward enough, but it may not always be practical. Suppose the manufacturing floor produces 400 different kinds of end products. How many different products (SKU, or stock-keeping units) are sold in a given day? Perhaps as many as 100 different SKUs might have experienced the sale of at least one unit. Most SKUs sold in a typical day are in the top 20 percent of products sold, and that group is likely to sell almost every day.

Does it make sense to generate up to 100 new production orders every day, when some of these orders will be for only one unit, or perhaps just a very few? This kind of policy seems to encourage inefficiency. It could mean that with an aver-

---

* The concept of aggregating demand from multiple distribution nodes at a single point will be addressed in greater detail in Chapter 8.

age production time of 10 days, at any given time, ten different production orders might be open for the same item.

## The Production Planning Conflict: Large Versus Small Batches

Let's view again the logic at work in such an environment (Figure 6.5). In order to maintain availability of all stocked items all the time, we must ensure the fastest possible replenishment time. To do that, we are driven to generate new, small production orders daily. But in order to maintain full availability, we also must ensure adequate protective capacity on our most-constrained resource. And to do that, we should complete production in batches at less frequent intervals. On the one hand, we are under pressure to generate new, small, production orders daily. On the other hand, we are pressured to consolidate like orders into larger batches. Obviously, we can't do both.

As sales continually deplete finished stock, quick replenishment is required to keep the volume of inventory, both finished and work-in-process (WIP), from getting out of hand. And production of fewer items per batch takes less time than production of more items, so small batches completed at more frequent intervals would seem to support faster inventory replenishment.

However, smaller, more frequent batches also imply more setup changes for different products. Unrestrained setup time, especially at a CCR, can quickly eat up the protective capacity required to accommodate unexpected demand surges. A CCR could easily become a bottleneck, which could create more—not fewer—stock-outs.

The conflict depicted in Figure 6.5 illustrates this dilemma, but it also suggests possible solutions. We need to ensure the integrity of both requirements, but how can we satisfy both of those needs? We must find an acceptable middle ground position.

## *Solution 1*

We retain Prerequisite #2, but find an alternative way to safeguard the replenishment time. How is this possible? One way is to determine some minimum required batch size to justify a new production order. For example, if the inventory target level is 100 units, and only one is drawn from that number, we now have 99 in stock. Rather than create a new work order for just one item, we will order production of a minimum batch size, say 30 units. One of those 30 will backfill the unit consumed, the other 29 units will increase the overall inventory by some amount above the target level. Continuing this way ensures that actual stock will normally fluctuate between the target level and target level plus predetermined batch size. The maximum inventory will always be:

Departure-Target + Minimum Batch − 1

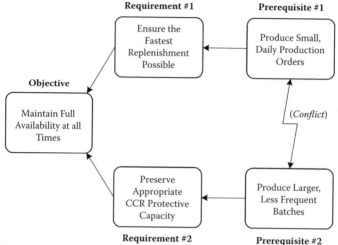

**Assumptions:**

1. We want to maintain the minimum required finished inventory.
2. Small finished inventories must be replenished quickly and frequently.

**Assumptions:**

5. Large inventory "holes" take longer to fill.
6. Large "holes" result from accumulating demand.
7. Smaller, daily batches maximize quick, dependable replenishment.

Requirement #1

Ensure the Fastest Replenishment Possible

Prerequisite #1

Produce Small, Daily Production Orders

**Objective**

Maintain Full Availability at all Times

*(Conflict)*

Preserve Appropriate CCR Protective Capacity

Produce Larger, Less Frequent Batches

Requirement #2

Prerequisite #2

**Assumptions:**

3. We require protective CCR capacity to respond to unexpected surges in demand.
4. Fast response to unexpected demand helps safeguard availability.

**Assumptions:**

8. Setups can consume a significant amount of CCR capacity.
9. Smaller, more frequent orders require more setup changes.
10. Larger, less frequent batches conserve setup time.

**Figure 6.5  Large or Small Batches.**

So whenever the actual stock of an "availability" item in finished inventory and production combined, falls short of 100, a production order for the minimum batch size is immediately created and released. If the minimum batch size is not sufficient to replenish inventory to the target level of 100, the batch is increased to the point that it does. Most production orders, however, will be released for the minimum batch size. In our example, above, the total stock in the system will normally fluctuate between the target level (100) and the target level plus minimum batch minus one (129).

This scheme should ensure a minimum batch that maintains a total setup time low enough to prevent overloading of resources. After all, this is *the reason* for dictating a minimum batch, unless the concern of creating an artificial bottleneck leads us to prefer that no batching occur. When we find a particular item for which the average daily sales are below the minimum batch, production orders won't be generated every day, but the replenishment time won't be too seriously compromised, either. Instead, we pay an acceptable price in the form of more stock, on average, held in the system—somewhere between the target level and the target level plus a batch.

## Solution 2

The drawback to Solution 1 is that it adds inventory that is not required to protect product availability. A second possible solution considers the likelihood that a new production order might wait too long on the production floor for CCR processing. Thus we could delay the release of that order. When the order is finally released, its quantity will be somewhat larger. How can we know ahead of time that it will be quite some time before the order is processed by the CCR? The planned load value provides that answer. The orders included in the planned load are only those that have been released to the production floor. Thus, at the time we are contemplating releasing a new production order, we will have a fairly accurate idea of how long that order would have to wait, on average, until the CCR is able to work on it.

If the time required to reach the CCR is much shorter than the expected wait time, there's no point in releasing that production order today. By not releasing such orders immediately, the accumulated amount of WIP inventory in the system decreases. However, the replenishment time isn't actually affected because an immediate release wouldn't reduce that time anyway; it still has to wait to pass through the CCR. So we might delay release of that order until the next day. Meanwhile, the CCR proceeds to process the orders already in production, shrinking the planned load as this work is completed. During this same time, additional sales may have cut into finished inventory even more, so the replenishment quantity would increase, and this increase could be added to the unreleased order.

## How Many Production Orders to Release?

How many new production orders can we release without incurring too much waiting time? In deciding what orders to release, the relative priorities for the competing items and the state of planned load guide our decision. For one production order to assume a higher priority than another, its inventory buffer must be more deeply penetrated, making the danger of a stock-out more immediate. This established the priority sequence. We then "draw the line" below the last order that pushes the planned load to approximately 80 percent of the estimated production lead time.

Here's an example (Figure 6.6). Suppose that the replenishment time is, on average, 12 days, the planned load is 160 hours of work, and the CCR works two shifts per day (16 hours). This means that the current planned load is about 80 percent of the average replenishment time. The work already in process will occupy the CCR for the next 10 days. The time it takes an order to reach the CCR cannot be more than 80 percent of the total production time. (Actually we assume 50 percent of the total production time is more than enough to reach the CCR.) So we can be reasonably sure in most cases that an order will arrive at the CCR before its turn to be processed.

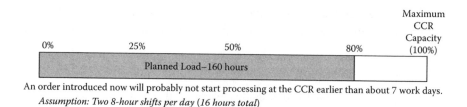

An order introduced now will probably not start processing at the CCR earlier than about 7 work days. *Assumption: Two 8-hour shifts per day (16 hours total)*

---

**Figure 6.6   Predicting Waiting Time Using Planned Load.**

A prudent decision might be "release orders to maintain planned load at 10 days or less." As operations at the CCR shrink, the planned load diminishes. New work time for new orders is released according to the needs established by the seriousness of the "holes" in finished inventory plus the WIP. However, as soon as a release pushes planned loads above 160 hours (80 percent of the replenishment time), new releases will be withheld until the planned load drops below that value again.

Suppose that tomorrow, before the decisions on releasing new orders are made, the total planned load on the CCR is 148 hours. We know that it's acceptable to release new production orders that require 12 hours from the CCR (including the setup time). Now, if we have 24 different items that need replenishment of various quantities, and the total load of those on the CCR is 20 hours, it means some of those orders won't be released today.

Let's assume we defer for a day eight orders whose cumulative load on the CCR amounts to 8 hours. While waiting that additional day, new sales decrease the finished inventory for those items. Now, before we even release those deferred orders, the incremental increase in planned load represented by those items increases to 11 hours instead of 8.

The increase is relatively small because each order, though for larger quantities, requires only one setup. Now, however, several other items need replenishing. Let's say that the load on the CCR is decreased to 145 hours during the intervening day, but the 15 hours available the next day is not enough to process all orders and still remain at a planned load of 160. Again, priority decisions must be made. Though we may not be able to accommodate all eight deferred items the second day (some of the new orders may have precedence based on high sales in the previous day), the chances of including specific items delayed from yesterday in today's release are still very good.

The primary advantage of Solution 2 is that it does not assume a fixed minimum batch at all times. At peak-demand time, orders might wait several days, accumulating higher quantities to replenish, so the actual batch released would not be very small. In off-peak times, each day's sales might be converted immediately to production orders without overloading the CCR. If this is possible, why not do it?

There is an important assumption underlying Solution 2: CCR setup time is significant enough to require consideration. In other words, failure to account for it, yet accepting small batches for processing, can cause a bottleneck to emerge.

When this assumption is valid, smoothing the load by matching the batches to the CCR's planned load can be effective. However, another work center having a high setup time could potentially cause the emergence of a different bottleneck. In this case, enforcing a minimum batch is a must. Besides minimum batch sizes, monitoring the load from released orders on the CCR is an effective way to refine batch sizes so as to match the actual batches to the CCR's pace without increasing replenishment time.

## Step 3: Manage the Buffer

Buffer management enables the effectiveness of making to availability. Production orders are planned without due dates. Operators on the shop floor need to process and move any work order that appears at their site, but they must also know which of several to do first. Consequently, a priority policy is essential. The operators, as well as the production manager, must know whether extraordinary effort is required to process and move a specific order. Even when no special effort is required, operators must know which order to begin next. The role of buffer management is to provide the appropriate priorities and to signal operators when extraordinary effort is required to move particular orders forward.

Let's quickly review the role of buffer management. When we insert a buffer to protect our production plan against variation, the consumption of that buffer indicates the current status of risk to the protected area. When too much of the protection has been consumed, the danger of exhausting all of it and compromising the plan is real. If this happens, extraordinary measures are required to maintain the protected function intact.

Unlike making to order, in MTS the form of protection used is *stock*, not time. When demand increases, protection decreases very quickly. The protection remaining is represented by the completed units still available *and* by the units on the way that we expect will arrive before finished inventory is completely exhausted. If an emergency signal were to appear when a minimum stock threshold is penetrated, the longest-standing production order for that item could be expedited to prevent the shortage.

Depth of penetration into the buffer determines priorities. Whichever item has the highest buffer penetration (i.e., the least remaining buffer) earns the highest priority. Buffer status at any point in time is the ratio of missing units to the target inventory level, expressed as a percentage. Pipeline orders are the older production orders that lie downstream from an order whose priority is being determined. The formula is:

$$\text{Buffer Status (\%)} = [(\text{Target Inv level} - \text{On-hand Inv} - \text{Pipeline Orders}) \, / \, \text{Target Inv Level}] \times 100.$$

Consider the following example for examining buffer status of a product with a Target Inventory Level of 200 units, but only 104 items on hand. Note that we are evaluating status for a product, not the buffer status of a production order. The buffer status is $[(200 - 104) / 200] \times 100 = 48$ percent, indicating a *yellow* condition and not prompting immediate action for this product. Now let's consider a production order (Order 1) for 44 units and assume it is the oldest production order not yet released. As we look at this order, its size is immaterial. Its purpose is to protect commitments to customers from the risk that the finished goods inventory might not be enough to satisfy demand until this order arrives. Thus, the most important consideration is how urgent it is to avoid a stock-out, regardless of the order size. Hence, the buffer status will still be 48 percent—exactly the same as the product itself. This order has a *yellow* priority when it competes with other products for the capacity of the CCR or any other work center that might have a queue of orders.

The production order for 44 units (assuming no other production order for this specific product) is released, meaning that in the system on-hand plus pipeline yields a total of 104 + 44 or 148 units—still 52 units below the target inventory level. In fact, there may already be some plans to produce more of this item, but the orders have not yet been released. Now assume a new production order, Order 2, arrives for 52 units; what priority should it receive? Its buffer status measure is the $[(200 - 104 - 44) / 200] \times 100 = 26$ percent. This means that Order 2 is assigned a *green* priority when it enters into the pipeline.

Notice that the priority is associated with replenishing the inventory to the target level, given that there is competition with other products for the CCR's capacity. When Order 2 (for 52 units) shows up some time after Order 1 is considered, the buffer status for that order is *green* $[(200 - 104 - 44) / 200] \times 100 = 26$ percent, meaning that its priority is lower than Order 1. When Orders 1 and 2 compete for the CCR the *yellow* order takes priority over the *green* order. Certainly when the CCR queue has a production order for another product whose buffer status is 75 percent (*red* status), the CCR would work on it before either Order 1 or Order 2.

The manager or operator has some flexibility beyond strict adherence to the buffer status percentage when selecting the order to process next. For example, when one order has a 45 percent status and another has a 42 percent status, other factors may enter into making the priority decision. While the order with 45 percent status has a slight edge over the 42 percent order, an opportunity to save time for a significant setup could be reason enough to put the job with the 42 percent status first. Neither order is required immediately, so surely we can consider other relevant factors without being forced to strictly observe buffer status.

On the other hand, if an order has a status figure of 85 percent and is competing with one that has a 35 percent buffer status, clearly the *red* (85 percent buffer penetration) order represents an "I am urgent" signal that should not be ignored. We strongly recommend using the priority scheme arising from the *red, yellow, green* buffer structure in exactly the same way they are employed in a MTO environment. *Red* condition orders often demand management intervention to ensure that any necessary action is taken to complete the order and avoid a stock-out. Otherwise, the typical shop floor guidelines for reacting to buffer status provide the necessary foundation for successful inventory management.

## Step 4: Maintain the Correct Target Levels

The finished inventory target level is initially based on a combination of real data and intuition, *especially intuition*. Either could prove wrong because of misreading the situation or because the situation has changed over time. If we are wrong, we must revise our situation depending on the direction of our error.

What happens when the existing target level is too low? The most immediate indicator is that finished inventory will decrease to the point of penetrating the red zone. A stock-out won't necessarily occur because when the red zone is penetrated, it prompts managers and supervisors to take action to expedite orders that usually are already on the floor. Even with the correct target level, red zone penetrations are bound to happen, but these should be irregular and infrequent.

However, if the target level is significantly lower than it should be, red zone penetrations will be more frequent and take longer to resolve. If you experience this condition, increase the target level.

What happens when the existing target level is too high? The most obvious effect is too much finished stock on hand. What qualifies it as "too much"? When the finished inventory status spends too much time in a green condition, there's more finished inventory than required (Figure 6.7).

A target inventory level that hovers consistently in the yellow-to-green zones is just about right. When actual finished stock is in the yellow zone, we expect enough additional stock (replenishment to the target level) to be somewhere on the shop floor. Thus, even if sales are surprisingly brisk, more finished units will arrive fairly quickly. Inventory being in the yellow zone is considered "safe" for meeting the availability commitment. However, being in the green zone is "too safe." Temporary excursions into the green are acceptable, but if the inventory level stays green too long, it's a signal that the target level is too high, so reduce it. Remember: a truly effective, efficient make-to-stock process is one that runs close, but safely so, to the ragged edge of danger. The assured availability represents effectiveness; holding finished stock to the absolute minimum required for assured safety represents efficiency.

For now, we will defer discussion of the detailed rules for identifying situations requiring a change in the target level and by how much. This discussion is more germane to distribution systems (Chapter 8). The problem of setting the target levels is more relevant there.

## Load versus Capacity in Inventory Management

In a production facility a requirement to raise the target inventory level has another critical side—the load versus capacity level. What happens when a lengthy peak load

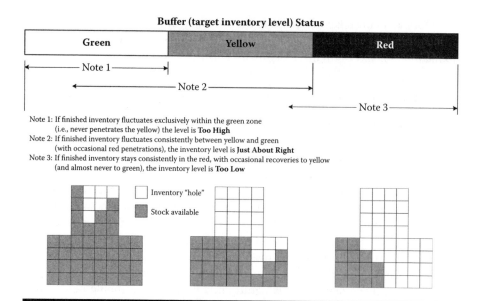

**Figure 6.7   Target Inventory Level: How Much? (1).**

turns the CCR into a temporary bottleneck? Remember, production lead time is *highly* affected by the amount of excess capacity at the weakest link (the CCR). Therefore, when the CCR is fully loaded, the production lead time stretches out disproportionately.

In making to availability, this will normally cause multiple inventory items to penetrate the red zone. When this happens, it becomes difficult to expedite *all* the affected orders very much. Moreover, while the CCR and other resources are busy processing red-zone orders, what will happen to the yellow orders waiting for processing? In a short time they, too, will slip into the red (sales continue to take place). Thus, the high load on the CCR causes many items to be in the red for what might seem a long time. Naturally, this sends a signal to increase target levels for a larger number of items.

Now, what happens if we decide to increase the target levels of some items? Let's say an item has a target level of 100 units, with a red level of less than 34 units. Now suppose the current level is 12 units, and the actual inventory for this item remains in the red zone for more than 10 days. A signal to increase the buffer is received, and the decision is to increase the target level to 133 units. But the actual result is that a production order for an additional 33 units is generated. Of course, this order is *on top* of having to replenish all the other items. Unavoidably, releasing the production order to try to meet the increased target level adds even more load to an already overloaded system. The production lead time increases even more, causing still more finished inventory items to penetrate the red zone and remain there for a long time. This is a vicious circle—a negative reinforcing loop—of increasing target levels and fighting shortages all over the place (Figure 6.8).

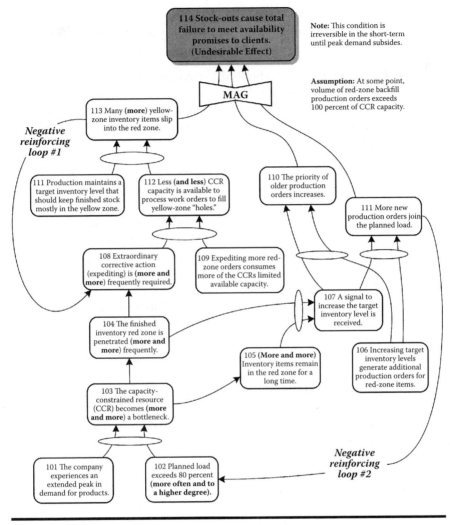

**Figure 6.8   Target Inventory Level: How Much? (2).**

The important point is not to fall into this trap. *Before increasing any target level in the shop floor, make sure you have adequate capacity for it.* Monitoring for capacity making to availability is discussed below.

## Managing Stocks of Components

Until now, we have focused our attention on making to availability for finished goods, a commitment made to certain clients. But what about components that are common to many different products we produce?

Whether finished items are made to order or to stock, the important consideration is whether or not many of those items use the same component. When they do, there are two ways to manage component production:

1. *Making to order.* Each time a finished item is required, the appropriate quantity of a common component would naturally be part of the production order.
2. *Making to availability.* In this case, the common commitment is not intended for external clients, but for whichever internal production line requires that component.

When is it beneficial to manage the common component to availability? The most important benefits accrue when production lead time for finished goods can be significantly shortened by having the common component available in stock instead of having to produce it from scratch. Sometimes this lead time can't be reduced. If the finished item contains other components that are *not* managed to availability, and that part of the process takes the same time (or more) to complete than our make-to-availability component, then MTA won't yield a real benefit.

Even if the production lead time of the finished item can be significantly reduced due to producing the components to availability, will the company as a whole actually benefit? In other words, can the reduced production lead time help in leveraging more sales, or facilitate a better selling price? If so, then making to availability is clearly the sensible approach. Moreover, when finished items (as well as the common components) are managed to availability, there's a potentially significant benefit in the ability to use lower target inventory levels.

Once the benefits for producing the component to availability are established, the cost of the move should be evaluated; in other words, the trade-off in cost of holding an inventory of completed components before a real need for them materializes. The higher the demand for items that use common components, and the higher the number of finished items that require the same component, the less costly the decision to manage that component to availability.

Once the decision to produce component X to availability is taken, avoid the temptation to start by determining the required target level immediately. Rather, expect that the buffer already provides availability. The basic stock must be in place before announcing X as a stocked item. This added stock must be produced *on top* of the routine demand for component X generated from firm orders for the finished items. *Build the component stock with an eye toward the amount of existing excess CCR capacity.* Whatever you do, *don't* make such a change at peak demand time.

Managing a component to availbility is essentially the same as managing a finished item to availbility. A target component inventory level is defined, production orders to replenish the stock are generated, and buffer management is put in place. From the execution point of view, the buffer's status when an order is received provides the relevant information for determining priority. There's no need to distinguish between production for finished items and production orders for components; they are all equal, as long as the green, yellow, and red priorities are followed. When a component is designated a stocked item, meaning that it is included to ensure availability, the management of finished items considers the component as raw material that should be available for any need at any time.

## Mixed Environments: Managing both MTO and MTA

Making to availability is a very serious commitment to potential clients. A sudden change in the market's preferences or needs could result in writing off some inventory, possibly a significant amount. It should be intuitive to distinguish between fast and slow movers. A fast mover sells well at all times, and consequently demand fluctuations are not very large. A slow item, on the other hand, sells only sporadically, and the sale quantities might vary. Demand fluctuations for a slow mover are likely to be relatively very large. Because the inventory target level represents the *maximum* sales within replenishment time, and because that maximum for a slow mover is much higher than the average sales, the target inventory level for a slow mover will seem relatively high. Therefore, the cost of making a slow-moving product to stock will be much higher than the cost for a fast mover.

Clearly, it would be preferable to offer slow movers as make-to-order items, basing the offer to the customer on delivery reliability and an acceptable response time. This confines availability commitments to more appropriate items, while not reducing the variety of products offered to the market.

Thus, the average manufacturing environment is likely to have a mix of products, some of them maintained for availability and the rest manufactured strictly to order. All of both categories of items require the same resources. Because we use S-DBR, we don't schedule the CCR, for either make-to-order or make-to-availability items. There are no differences here.

How would we use buffer management to establish the relative priorities between making to order and making to availability? One type of order has a specific due date, while the other does not. Does this mean that the order with the due date to a customer (make-to-order) will always have higher priority than an order for availability? Not necessarily.

The buffers are obviously different—one is expressed as *time* buffer, the other as stock. Yet the meaning of buffer penetration is exactly the same, so the colors mean the same thing. A red order is a red order, no matter which type of buffer is penetrated. It doesn't even matter whether the red item is intended for component stock or as an end item.

## When a Specific Product Sells both from Stock and to Order

Traditional manufacturing practice does not clearly distinguish between items made only to order and items made exclusively for stock (or availability). Manufacturing resource planning (MRP) admits mixing the two in the master production schedule (MPS).

The TOC, however, *does* make the distinction. When a customer asks for N units of Product M to be shipped at a certain date, the obvious buffer is time. When two or more customer orders share the same due date and have the same time buffer size, it would be acceptable to merge them into a single production order. TOC would allow adding units for stock only in those cases where the use of a minimum batch size is absolutely necessary. Thus, in TOC, except for minimum-batch situations, we distinguish between orders to be delivered to a client by a firm date and orders that replenish consumed stock.

Does this mean that *every* finished item must be either for order or for stock? In principle, the answer is yes. Only the characteristics of the item and the protection mechanism (time or stock) are different. It's much simpler and more straightforward to treat each item as either for stock (availability) or to order. In reality, though, there are situations in which an item can be *both* at the same time, depending on clients' needs. There are two basic scenarios in which this might happen:

1. *Make-to-order for a relatively slow-moving item.* This is the typical approach for slow-moving products. But sometimes clients need an item immediately and are ready to pay a premium to get it *now*. For example, let's assume that the item is a critical spare part for a large system. Demand for this kind of item would be sporadic at best, so making to order would be the preferred approach for such items. However, occasionally when a need arises and the client has a major project or operation on hold for lack of such a spare part, a really good, reliable supplier should be able to provide one or two units from stock immediately even though the item would almost always be made-to-order. Of course, this kind of service ("insurance") ought to be handsomely compensated. But doing so means holding a very small stock of normally made-to-order items just for such occasions. In all other situations, such items would be produced to order because the need would not be very urgent.

2. *Make-to-stock for a known demand.* More commonly, a fast-moving item is made to availability, but a reasonably firm demand is known ahead of time. Some clients simply know what their needs are in advance and advise the supplier accordingly. These clients don't care whether the supplier provides them from stock or produces it specifically for them. All they care about is that all required items are in their hands at the specified date.

The complexity in these two situations derives from the use of two different buffers for the same item. One suggested simple approach is to split the item into two different SKUs. One would be exclusively for make-to-order items, the other for managing for availability. This poses a dilemma between simple solution and a sophisticated one (Figure 6.9). A more complicated solution might add an incremental "delta" to overall performance, but it might also put successful implementation at risk.

It seems to us that the question of make-to-order versus make-to-stock for a product with both types of demand (immediate availability versus "I need it at this date") is an excellent example of the simple versus complex dilemma. It's certainly possible to find a combination of the two that would serve a particular case somewhat better. Visualize a make-to-order job stuck in production while there's enough finished stock to ensure timely delivery. But because of the separation between the items, there is a need to expedite that order.

How much complexity are you prepared to introduce into a process to overcome such an obstacle? How do you verify that such a process will *always* yield the best results? What implementation risks would be exposed if you try to teach everyone in operations how to deal with such a situation?

Suppose the simpler process is in place, prompting the production manager to intervene when a make-to-order job is in the red zone of the *time* buffer. The production manager probably knows the products and the process well, so he's aware that the make-to-order item is identical to the make-to-availability one. Though the production software treats the two items as distinctly different SKUs, the human manager would know that they are the same. Then, instead of expediting the make-to-order job, the production manager might prepare the shipment from stock and route the make-to-order job to inventory to replenish the stock he used. Instead of

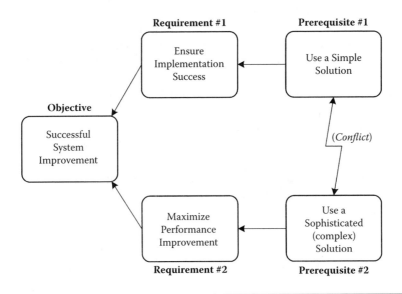

**Figure 6.9   Capacity Overload in Making to Stock: A Negative Branch.**

making the normal process more sophisticated, we depend on the human manager to take the appropriate, more sophisticated action.

This, then, is what we recommend doing: implement the simple rule that an item can be made either to order or to stock, and only if necessary create two apparently different items. Leave some exceptions to be treated manually.

# MTA and Capacity

One of the prime functions in using S-DBR for making to order is determining the "safe" dates that we might offer customers, based on the planned load of the CCR. Tying sales and production jointly to one clear element of information relevant to both makes it possible to "squeeze" the CCR's capacity without compromising reliability. When market demand rises, safety dates are pushed farther into the future, unless the company quickly intervenes to add capacity. This ability to smooth the load on the CCR by promising due dates that are in line with capacity is missing in making to availability.

The most threatening factor in committing to full availability is a sudden lack of capacity. As with time buffers in make-to-order, the stock buffer in make-to-availability has a short-term effect. When the total load on the shop rises, existing target inventory levels provide some protection, but only for a short time. If replenishment is very slow, there's no longer any assurance that shortages won't occur. When capacity is constrained, it may already be too late to increase stock buffers, and, as previously discussed, increasing stock buffers in the face of zero spare capacity would actually worsen the situation.

There is only one way to eliminate this threat: maintain sufficient protective capacity and monitor it very carefully so that when planned load exceeds the safety level (approximately 80 percent of the production lead-time) there's still time to add capacity or to rein in market demand.

In this chapter, we have identified a need for a number of firms to protect critical resource capacity by augmenting MTO with MTA for a select set of products. In addition to pointing out the principles that may lead a firm to do this, we have outlined a process for implementing the hybrid system that is primarily MTO with an added portion of MTA to enhance the availability of the standard products for our customers. In the next chapter, we will examine some of the potential hurdles that may arise for make-to-availability and situations that look more like make-to-stock.

*Chapter 7*

# Make-to-Availability, Make-to-Stock, and Make-to-Order Issues

## Contents

The basic make-to-availability (MTA) approach that we introduced in the previous chapter is strikingly different from making to order (MTO). Realizing this, we must consider even more possible complications in managing it. Shortly, we will address seasonality, a topic we will also cover in distribution. Another issue is dependent setups when they are performed primarily for availability. We will also discuss certain aspects of making to stock that don't necessarily involve a commitment to availability, such as peaks of very short demand. These pose a different challenge because there's no practical way to replenish within the peak—one had better be right the first time.

Another complicating and interesting topic is making to stock in a situation where the client's tolerance for delivery is more than zero (immediate), but still shorter than production-plus-transportation time. In this particular case, we are really in a MTO situation in which the standard lead time is too short to protect on-time deliveries.

Let's begin where we left off in Chapter 6, considering load-capacity relationships. But now we will expand the discussion to include a mixed environment of MTA and MTO.

## Monitoring Capacity in Mixed MTO and MTA Environments

What happens when a company is required to both make to availability and make to order at the same time? This means that certain products are produced and delivered by demand/discrete orders while others are fulfilled from finished stock. We've already seen that buffer management gives us effective priorities, thus managing MTO and MTA together in the execution phase is the same as managing MTA alone. The real challenge lies in planning: how do we assure that we have enough protective capacity to accommodate both?

When we practice MTO using safe dates (based on the planned load, plus half the time buffer), we have an effective procedure for smoothing the load. Consequently, any increase in demand will extend delivery time quotations. Some clients might not be satisfied with such a delayed promised delivery date. They might look for providers elsewhere. When such a threat to our market is significant, we should seriously consider increasing capacity. However, that option must be carefully controlled to ensure that reliability won't collapse.

In making to availability, tighter load control is required, as we have already seen. That additional load control must ensure that we retain enough protective capacity to compensate for Murphy's law and any sudden shift of priorities. To ensure that adequate protective capacity is available, the total demand must be kept below a certain level. Because it's not possible to be sure that the total demand won't suddenly exceed the desired limit, we monitor the planned load.

We are then able to identify the potential loss of the protective capacity well in advance of it actually happening. We depend on marketing and sales efforts to sustain market demand at a defined level below the limit imposed by protective capacity. If the planned load approaches the protective capacity limit, management must be prepared either to increase capacity or to apply some kind of rapid restraint on the market.

## Three Scenarios

There are three possible scenarios that must be considered:

1. MTA requirements are much larger than the MTO requirements. For example, the MTO part of production consumes less than 33 percent of the CCR's total capacity.
2. MTO is by far the larger part.
3. MTA and MTO are about equal, each requiring more than 33 percent of the CCR's capacity.

## MTA Greater Than MTO

In this situation, MTA is the primary focus of our attention. MTO requirements, though present, are the exception rather than the rule. This situation is actually quite common. When manufacturing companies produce standard items for which MTA is appropriate, these items can quickly become the main market. Yet the clients for those standard items may also want some nonstandard items. These would naturally be slow movers. In order to maintain a complete competitive edge, a company must provide those slow items, but they are not likely to become a particularly large part of production output. Those slow-movers would best be provided as MTO items.

When the MTO part of production is less than one-third of total capacity, the larger MTA component already forces us to ensure availability of more than enough protective capacity. The combined planned load should contain the following loads:

- All MTO orders that must be shipped within the standard lead time.
- All MTA orders that have been released, plus all replenishment requests that have not yet been released.

There is no practical way to use the safe-date determination mechanism for the MTO orders, meaning we can't just add half the buffer to the planned load to get a safe date. The problem is that if you base your safe date on the existing planned load and promise delivery at what looks like a safe date, an increase in the sales of MTA items would cause the release of production orders not accounted for in the planned load when the MTO commitment was made. Moreover, increased sales could shift priorities leading to expediting of several MTA items—again not considered for when MTO promised dates were set. It doesn't make sense to use the technique of reserving capacity for MTA when the reservation is for more than half the available capacity. The uncertainty regarding the actual use of the reserved capacity is too large.

Thus, the appropriate safety mechanism for the relatively few MTO orders is to quote the regular standard lead time and validate that there is enough protective capacity to allow the priorities to keep all commitments intact. Rapid response orders can be used only in the presence of an even higher level of protective capacity. The problem is that *there isn't a reliable way to determine the "right" level of protective capacity* a priori. Eventually, the experience gained from observing the ratio of red orders to open released orders helps identify when the planned load is too large. When the number of red-zone orders goes up considerably, it signifies that the load is close to the limit.

One pitfall of which we must be aware is that the normal MTA horizon is the average replenishment time, which is normally *shorter* than the standard lead time for MTO in the market. Because we are considering MTA the dominant environment under this scenario, we should emphasize replenishment time as the baseline for assessing the state of available capacity.

## MTO Greater Than MTA

In the second case, at least two-thirds of the demand is for MTO. In this situation, we should assuredly attempt to smooth the load by quoting longer delivery dates when CCR capacity is constrained. Still, the lesser amount of MTA requires adequate protective capacity at all times, so that it won't consume too much capacity needed for the MTO component or cause delays, or stock-outs of MTA items. So the obligation to provide MTA items precludes us from simply quoting safe dates. At the time a delivery date is promised to a MTO client, we don't know how many urgent MTA orders might be initiated and require completion before the MTO delivery date to which we just committed.

Reserving a certain level of capacity to deal with the MTA orders, especially those that would require expediting, can be useful here. Capacity can be reserved in two different ways:

1. *Reserving the full capacity MTA requires.* If the MTA load averages about 30 percent of the total output, let's add another 5 percent (total of 35 percent) to ensure enough protective capacity. This means that MTO safe dates would be based on their own (separate) planned load, which includes *only* MTO orders. However, this planned load considers only 65 percent of the total capacity to be available. Now, when the MTA load grows, it will first consume the 35 percent of its own capacity allocation before consuming any of the MTO

capacity, which might cause MTO orders to penetrate their own time buffers. The possible pitfall here is that separating MTO from MTA is never perfect. Overflow of MTA capacity consumption might not be noticed until too late. Naturally, the more capacity reserved for MTA, the less the risk. But that also means more possible limitations on the sales of MTO products, or having to promise later dates.

2. *Reserving capacity selectively.* The first approach is conservative in the sense that it might set aside too much capacity for MTA items. A more integrated approach would be to reserve capacity only for the estimated volume of MTA orders that would be released after the last current MTO delivery has been determined. Then some of those orders would be expedited. In this case, the new, unanticipated MTA load would take priority over already-accepted MTO orders because their penetration into the buffer is higher than most of the MTO orders. Most MTA orders don't require special expediting efforts. As long as on-hand finished-goods stock is in the yellow zone (and certainly if it is in the green), a particular MTA order isn't really urgent. And if sales for that particular item are currently slow, the production for it can be delayed. Consequently, many of MTA orders require longer time than the average replenishment time. So not too many MTA orders released after the MTO safe date is determined would be completed ahead of the MTO order. For example, let's assume that 50 percent of MTA orders (not yet released) would eventually require completion before the MTO order. In this situation, reserving capacity equivalent to *half of the ratio of MTA to the total output* would be sufficient. If MTA represents about 30 percent of total output, reserving 15 percent should be adequate. The MTO safe-date determination must be based not only on the MTO orders already committed, but also on MTA orders already released. Thus, the planned load would now be based on 85 percent of total capacity (allowing for the 15 percent previously reserved for urgent MTA orders). But it includes MTO orders and all released MTA orders.

Some clarification of the planned load concept and how it's used is in order. The simple idea behind it is adding *all known demand within the designated time horizon* to get a rough idea of the current load and how long it might take to process it all, assuming no idle time. We have already discussed various considerations concerning the time horizon. What we are considering now is what orders to include: only MTO orders or including the existing MTA orders? The answer to this question depends on what we want to do with the time horizon. In the first way of reserving capacity, the MTA orders were treated separately. They were given full capacity reservation. In the second way, we've integrated the existing MTA into the main planned load, reserving only optional capacity for MTA orders that have not been released, but which might be released shortly and perhaps even expedited.

## MTA Is Approximately Equal to MTO

It's not very common for a company to experience nearly equal demand on capacity for MTO and MTA items. Confusion surrounding which orders are MTO and which are MTA can mislead management to conclude that the relative impact of MTA and MTO on capacity are about equal in many cases. Historically, production work orders have been a mix of production for firm orders and for stock. The stock is not designed to ensure availability, but rather to level the work orders to an "optimal" batch size and to provide an opportunity to ship from stock, when applicable, without committing to having items always available. Another case is the firm order, based on due date, supplied to clients who hold similar items in stock. This particular situation exposes an opportunity to offer to manage clients' stock for them—vendor-managed inventory (VMI)—and obtain a better overall deal. Practicing TOC in such circumstances would eventually lead the producer to make the majority of items to availability. But such shifts only happen over time, so it's possible that for some period of time MTO orders would require about the same capacity as MTA items.

Having about equal distribution of capacity between MTO and MTA requires the same level of protection as situations in which MTA generates the majority of total demand. MTO's advantage in smoothing the load can't be maintained when the MTA component is fairly large and fluctuates.

Thus, approximately equal loads in MTA and MTO are similar to the preceding scenario in which MTA dominates demand. We need to rely on maintaining enough protective capacity to ensure *both* availability and reliability. To validate that protective capacity is maintained, we use the planned load. The questions, though, are (1) What should be included in the planned load and (2) What planned load value should be designated to signal an overload?

The allowable planned load on the CCR must include all orders to be delivered within the chosen horizon. This horizon includes the longest standard lead time and the replenishment time that dictates the inventory level for the MTA items. The planned load assuredly encompasses all MTO and MTA orders already released. However, it also includes MTO orders that have not yet been released, but which need to be delivered within the horizon, *and* all items required to replenish all MTA items to their designated target levels.

Determining the reference line of planned load that, when crossed, would signal a possible lack of capacity. As long as the planned load does not take longer than 80 percent of replenishment time, and it is less than the shortest standard lead time by at least half the time buffer, you are certainly safe. It's common sense to expect the replenishment time to be shorter than the shortest standard lead time for MTO. This is because the typical MTA item is standard, while a typical MTO is not likely to be; it may need to be customized to the need of the client. Moreover, typical lead times for MTO include the protection required to account for Murphy's law, but replenishment time is an average. Protection is provided by the finished goods stock. Thus, when the planned load is only 80 percent of the replenishment time, protective capacity is sufficient.

However, it's clear that planned load could be greater than 80 percent of the replenishment time, and yet the combination of about 50 percent MTA and 50 percent of MTO with longer standard lead times still provides enough protective capacity. So how can we know the position of that imaginary line signifying the limit to which we can allow planned load to grow without risking availability and/ or reliability?

We have no formula for determining this line. We also suspect there can be no consistent formula when the ratio between MTA and MTO fluctuates and each market is, in itself, not evenly distributed in time because of our inability to quote safe dates. What's important to remember is that "reality signals you" whenever your production floor is under pressure. That clear signal is the number of red-zone orders. When it starts growing, it's a warning. So careful monitoring of the planned load and the number of red orders could give you an indication of whether your arrangement is still about right. However, there are disadvantages as well as advantages in pushing the system. By accepting more orders while monitoring the number of red-zone orders, as you see the red-zone orders grow, this is a clear warning signal. In reality, that rise may not be evenly graduated. You can be close to the limit and still not get the signal. But when you get it, *you'd better pay attention to it:* you are at the limit. You had better ensure some capacity buffer to relieve your immediate pressure. From that point on, you will have a reasonably good idea of the limit.

The notion of a capacity buffer is especially important for MTA, and even more so for a combination of MTA and MTO. Buffer management (time) provides a dependable signal of load pressure, but it might come too late if too many red-zone orders emerge at the same time. If maintaining perfect availability and reliability is important, you *must* have the option to increase capacity quickly, even at an undesirably high cost. Buffers are a way of life.

## *Dealing with Seasonality*

Seasonality means that you experience more demand at certain predictable periods of time. It can cause a certain kind of complexity. Part of it is artificial, but we need to examine some points carefully. Two important characteristics of seasonality can impact operations:

1. We *have* prior knowledge about an upcoming demand surge. Don't forget that buffer management can help identify a demand trend (upward or downward) and suggest a change in target levels. However, it can't predict a coming peak in sales. Until the season starts, a sales surge won't be detected. Then, when the season ends, buffer management may again be too slow to recognize the declining trend.
2. Capacity can be overrun. During the season itself, the CCR might actually

become a bottleneck. This means the stocks will be depleted and stock-outs will occur. The preferred way to handle this is to smooth the load on the CCR by anticipating what will be sold and produce this before the season begins.

Bear in mind that from the capacity point of view, it doesn't matter what items were produced to stock before the peak. Perhaps all the target levels were increased and properly replenished before the peak. Or perhaps some items have been produced to such high stock levels that sales would be covered throughout the whole season. The capacity saved is equivalent to the sum of CCR time to produce *before* the peak all items actually sold *during* the peak.

Seasonality is an anticipated demand peak in a defined period of time. Most common are seasons within a year, resulting either from climate changes or holidays, that encourage more sales during that period. But seasonality could span a quarter, a month (for example, the end-of-the-month syndrome), a week, or even a day. Restaurants often experience daily "seasonality." Where manufacturing is concerned, we confine our discussion to seasons of a year because their impact on production is especially noteworthy.

The primary debate concerning seasonality centers on whether to change inventory target levels in anticipation of the coming seasonal peak demand. The alternative is to leave it to buffer management to prompt a change, and then just make the change. In other words, to *react* rather than to *anticipate*. Why would we consider relying on buffer management? In other words, why should we ignore knowledge that we already have?

The question is, do we really know much about the peak? Most of the time all we really know is that a sales peak will occur. But can we know for sure which items will be affected and by how much? This is where forecasts creep in to seduce us with a number, such as "Sales will grow by 30 percent." The 30 percent figure is, of course, just an average based on past years' data. Three factors play a role here. The first is that it's just an average, and the real increase could differ significantly. Second, it could be specific-item dependent, meaning we could expect to see a much larger demand increase in some items, while others would increase substantially less—or not at all. Finally, there is the relationship between the demand and the target level. Suppose, for example, the market for a certain item surges by 25 percent. Should we increase the inventory target level by 25 percent? What about the replenishment time? Will it be the same "in season" as it is now?

Forecasting requires at least three years of documented results to reliably identify clear seasonality. In many markets, too many things can happen in three years. The behavior of consumers might change. In too many cases, all we know is that the season has an impact, but the size of that impact is hard to predict.

When so much is unknown, it might be best to wait for the peak itself, then increase or decrease the target inventory level based on the real, measurable pattern of demand as expressed during buffer management.

Nevertheless, we may let ourselves in for trouble if we ignore prior knowledge that a seasonal peak is coming:

■ The increase in the demand might be too high for the production floor to react quickly enough. Even when a good analysis based on buffer management prompts increasing the target level, it does not provide any indication of how much to increase it. E. M. Goldratt's general recommendation (2003) is to increase the level by 33 percent. But at the start of a peak this increase might be far from enough. More time would likely elapse before the repeated analysis shows a need to increase the target level again. Also, in order to increase the target level a degree of excess capacity is mandatory. This is not likely to be available within a peak of sales.

■ The onset of increased demand might be too fast relative to replenishment time. Suppose the demand goes up by 25 percent within the first day of the season, and the replenishment time is two weeks. On-hand stock would penetrate the red zone in a matter of few days, and stay in the red for quite a while because two weeks' worth of incoming orders are too small. The target level will be increased eventually, but it will take two weeks to stabilize the system after that buffer increase. That might be too late.

■ The most troubling point: Because the demand in such a peak usually encompasses many end-items, the unavoidable result is the loss of protective capacity, at least temporarily. This could extend replenishment time even longer. And longer replenishment times mean even higher target levels.

In manufacturing, when an overall demand increase of 20 percent or more is reliably predicted, preparation for that peak is essential. In many cases, a planned load increase of 20 percent for some extended period of time would inevitably penetrate protective capacity. We had better anticipate such a peak and be ready for it.

## *Increase in Target Levels*

While preparing for the peak, we still experience our normal demand. We realize that we must use our excess and protective capacity to increase the stock in the system in preparation for the season. Thus, the timing to start increasing the target levels, so that when the peak hits the stock levels will be suitable, depends on the level of excess-plus-protective capacity available before the peak.

We recommend planning the transition into the season very carefully. We should start by anticipating the required target levels during the assumed peak. To do that, we must predict the average demand and the average replenishment time during the season. Both are nontrivial assessments. On top of that, we must still incorporate a safety factor to safeguard both the demand (i.e., to be sure to meet maximum demand) and to avoid too long a replenishment time. By using such a broad factor, we reduce the compulsion to "accurately" predict demand and replenishment time.

Thus, we are likely better off in relying on intuitive prediction, augmented by a safety (paranoia) factor, because only imperfect information is available.

Once the season target levels are decided, we ought to build them up, carefully noting the time required to produce the increase in the target levels, while continuing to meet the normal ongoing demand. A gradual increase would be fittingly structured as a series of slow increases of all affected target levels, increasing a few of the items at a time and then moving on to the other items based on some priority scheme.

If peak demand during the season makes the CCR a real bottleneck, the replenishment time for some of the items will increase beyond the capability to restock everything within the peak. This means the stocks would go down considerably within the peak—too many items will be in the red; the priority system will be incapable of immediately guiding us to recover; and some shortages will occur. Additionally, this type of pressure usually leads to loss of capacity from breaking setups to help with a product that seems, at the moment, more urgent than another.

How can we accommodate peak demand seasons in which demand exceeds the capacity of the CCR? One way to handle this situation is to produce, before the peak, a very large quantity of several key items, well above their "official" target level at the peak of the season. This precludes having to replenish these items for quite some time, freeing the CCR to respond to the demand of the other items.

A decision to produce above the target level lies somewhere on the line between make-to-availability and traditional make-to-stock. On one hand, the ultimate objective is to provide full availability coupled with low inventory. In this case, we still want full availability, but the inventory costs are higher. Achieving both availability and low inventory, the TOC way is achieved by focusing on fast replenishment of the actual demand. Now on the other hand, facing a season that requires more capacity from the CCR than is available forces us to protect availability by freeing capacity and creating a larger than normal initial inventory. Not as efficient, but (from the market's standpoint) effective.

Certainly, demand for fast-moving items that are generally selling well tends to fluctuate less. The first task should be to determine the *minimum* quantity of those items that might be sold throughout the whole season (we assume that the shelf life of these products is not a factor). That quantity is relatively safe to produce. If those quantities, produced before the season, are sufficient to prevent the CCR from becoming a real bottleneck, then we will definitely profit from having them completed ahead of time.

Even though one might produce above the target level in anticipation of a peak season, it's critical to have an appropriate target level in place. If demand is high enough to reduce the initial built-up inventory down to where it penetrates the target level, that level will require replenishment, even if we are in the middle of the peak and the pressure on the CCR is high. After all, if we produce the minimum anticipated quantity to be sold throughout the season, in most cases, we could expect to produce somewhat more of specific items as sales rise above the minimum anticipated.

## Anticipating the End of Season

The end of the season involves one critical decision that has to be made: when to reduce inventory target levels back to their normal level, or perhaps somewhat less. When we anticipated the demand surge, we may have stocked up on selected items prior to the peak. When the peak surge has passed, demand will assuredly decrease. Waiting too long for demand to start down can be a major mistake because it can mean replenishing the current consumption to too high a level. If this happens based solely on peak pressure, our output may miss the season and remain on the shelf for a very long time. Consequently, anticipating the end of the peak is required, and a resulting gradual decrease in target levels would be a smart move. We want to start the drawdown of target levels about one replenishment time period before demand starts to deteriorate, if at all possible.

## Sequence-Dependent Setups in MTA

We discussed sequence-dependent setups in Chapter 5, when we analyzed the make-to-order environment. This tricky situation may introduce very high variability into replenishment time and leave us with almost no way to expedite (i.e., recover). This has serious ramifications for making to availability. Let's review the problem at hand. We are referring to a resource that has clearly a preferred sequence for processing a variety of the items to be produced. If you violate the preferred sequence, the amount of wasted capacity is very large indeed—possibly turning the constrained resource into a bottleneck. Also, when we maintain the preferred sequence, replenishment time could be equal to the entire cycle time (the time from producing a certain item through the production of all the other required items until the resource returns again to work on the same item). Depending on that length of time, availability may be compromised, because expediting requires changing the preferred sequence to the item that's most urgent now—and this can be done done only in very extreme cases.

Managing sequence-dependent setups requires much higher target levels than would otherwise be feasible. First, the replenishment time is longer than for similar shops that don't need to follow our preferred sequence. Second, because you can't rely on buffer management to dictate the priorities on the floor, you must maintain enough stock to protect against stock-outs during the processing sequence the constrained resource is required to follow. Sequence-dependent setups encourage making to stock because of the need to respond quickly in a system that can sometimes be very slow. Therefore, when you need it quickly, having stock might help.

One caution here is that the capacity problem exacerbated by doing more setups than necessary when one deviates from the sequence. If we decide to add more units to a batch within the sequence to avoid wasting capacity, we have succumbed to the temptation to make more than seems required (i.e., increased the batch size) at the time. We know the bad results that such a choice could produce. Making to availability is different in the sense that the commitment to the market is dominant. To meet our commitment, the costs we must bear are the higher stocks needed to protect availability for a complete cycle, thus reducing the need to alter the sequence and limiting expediting.

Buffer management in this situation serves mainly to determine when a change in the target level is required. This is not as clear-cut in a dependent setup situation as it is when the sequence can be easily changed without a huge impact on capacity. Because, by definition, resupply comes once in a cycle, the buffer status of each item continually deteriorates throughout the period. Thus, just before the next supply arrives, the buffer status is in its most urgent condition (inventory is lowest). Now would you like the inventory at just one minute before the supply arrives to remain above the red zone? Those who think the utopia of exact inventory levels is attainable would like the inventory, at the point of resupply, to be on the verge of a shortage. Thus, if it's in the red, but not out of stock, it doesn't seem to be all that bad.

Still, would we consider protection adequate if, over several cycles the inventory replenishment arrived just in time to prevent a stock-out? In short, the fact of only one resupply within replenishment time tends to cause target levels to be increased. It also causes us to continually question whether we need to increase the target levels even further.

From the operational perspective, it in a sequence-dependent setup environment is critical to update the quantities for each item at the very last minute. To the extent possible, each item should be produced, at most once, within a cycle. It's worth the extra effort to ensure that the updated number to be produced is determined at the last possible minute. The complexity of sequence-dependent setups for MTA is much less than for MTO. S-DBR implementation of MTA within such an environment is still straightforward. However, to truly ensure availability, the cost of the extra complexity is represented in significantly higher levels of inventory.

## Make-to-Stock That Is Not Make-to-Availability

Why would one decide to produce to stock without a clear commitment to availability? There are several possible reasons:

1. There is no easy way to offer availability—a certain quantity is to be offered, and when it's sold out, it won't be available again for quite a while. This situation is typical to very short but very high peak sales, such as a holiday, a short

exhibition, the first day a new Harry Potter book is introduced, or a specific brand of wine from a specific year. A certain quantity is produced and there's an obstacle to repeating production of exactly the same item. Even if it can be replicated, it might require an extended period of time before availability is reestablished. In these cases, one of the marketing points might be that availability is *not* guaranteed.

2. Use of capacity must be leveled throughout the year—as in the discussion above on seasonality.

3. Sometimes the commitment to the market is actually based on *time*, not on availability, but it's partially assisted by make-to-stock. A similar case is a make-to-order environment where the standard lead time is *shorter* than the required time buffer on the production floor.

4. Real make-to-stock that involves no commitment is sometimes required for exploiting the CCR. As is indicated in the previous chapter, MTA requires protective capacity. Is there a way to exploit that capacity without degrading the availability of any item that is included in the commitment to availability? This is possible only when the priorities are very clear: in no way will utilization of the protective capacity be at the expense of MTA. When such clear priorities are achieved, then making-to-stock of items in demand that carry no commitment whatsoever makes a lot of sense. Of course, one must validate that the items made-to-stock can be sold once they have been produced.

Let's examine some of the make-to-stock options. They represent a certain difficulty compared with MTA because decision making must consider more parameters. A better understanding of making to stock will also lead to a better understanding of the real meaning of making to availability.

## *Dealing with Very Large Short Peak Sales*

Sometimes replenishment time is longer than the duration of the peak, and the peak is short in duration. Once we know the actual demand, we may be unable to react to it in a timely way. The uncertainty regarding demand makes it unlikely that we can ensure availability unless we prepare substantial stock. When many SKUs are sold in a short but rather high peak, the investment in stock to ensure availability can be very large. If we were to attempt a commitment to full availability during the peak, the notion of a target inventory level is meaningless. There is no target to which we might replenish. There's only an uncertain target demand; the best we can do is to hope that it's sufficient.

A forecast is usually the entering argument in deciding the level of stock to prepare for such a peak. However, be *very clear* when asking about the type of forecast. The common forecast, which predicts the average demand (expected value) within the peak, provides little chance for an optimum decision. We must decide what to protect: product availability or the unsold inventory level after the peak is over. This is a critical marketing and economic decision, the considerations of which affect both the cost of the buffer used and the impact of being sold-out.

We suggest first obtaining a forecast of the *reasonable range* of demand for each item. The bottom of the range should represent the reasonable minimum that might be sold; the top of the range would be the reasonable maximum. Any rational decision should be biased toward one of the extremes of the range—the side that might avoid the greatest "damage." Thus, if selling out might compromise future sales and this damage is presumed to be greater than the possible penalty of having leftover stock, then the decision should be to produce the reasonable maximum.

Remember that the strength of TOC lies in its capability to replenish quickly. So when this seems impossible, please review the situation again. Is there a way to produce and distribute quickly enough within the peak? If this is workable, it represents a giant step toward maintaining acceptable availability with only a moderate (or even low) investment in inventory and capacity.

## MTO with Short Standard Lead Time

Suppose you are making to order, but the customer's expected delivery lead time is very short compared with your own production lead time (from material release through complete order delivery). This is a very common situation. And when delivering the order requires adding transportation time, local suppliers have a competitive edge. Customers don't care that one supplier has the goods in stock but needs to ship, while another supplier is nearby but doesn't carry the required items in stock. In neither case is delivery immediate, so a promised delivery time must be set. Suppliers who can commit to an earlier delivery could have a real competitive advantage. Thus, there's strong pressure to offer reliable but quick delivery.

Let's consider an example in which a competitive lead time is four weeks. The average reliable transportation time, depending on the distance of the client from the plant, is two weeks. The production time buffer is three weeks. This represents a reliable time from material release until the order is ready for shipping. In this situation, how can a company offer four-week delivery reliably to distant clients?

One option would be to make to availability at the plant warehouse. If the order lead time is just the transportation, this should work perfectly. But at what cost? The target inventory level should be based on the maximum demand within a three-week period (the length of the production buffer). This is somewhat high, considering that the tolerance of most clients is longer than just the transportation time alone (by two weeks).

The other option is to maintain only enough stock to cover for the difference between the required delivery lead time (four weeks) and the reliable production lead time (three weeks). About one week of stock could satisfactorily cover the difference. This option is considerably more complicated than either MTO or MTA because delivery reliability depends on combining of two buffers: a time buffer and a stock buffer.

To effectively manage the combination of two buffers, one must establish the proper priorities. First, stock must be built to cover the maximum demand for a one-week period (the difference between the five weeks for production and transportation and the client's tolerance time of four weeks). Let's assume that we already have the equivalent of one week's maximum sales located somewhere between the shop floor and the plant warehouse. This amount is not enough. Every day, orders are received to be delivered in four weeks.

To calculate a date when production should target completion of the order, we must deduct the reliable transportation time from the delivery date. Then the material release date is determined by subtracting the production time buffer from the target date for completion of production. If that release date is either today or in the future, this order doesn't need the stock buffer for protection. The time buffer is adequate.

However, orders with calculated release dates in the past really do need the stock buffer because their actual time buffer is shorter than required. Those orders are released immediately, relying on the probability that there will be enough stock (from that already within the system) to ship the appropriate quantity on the required date, if necessary. Once the production order is finished, it simply replenishes the stock buffer.

The inventory on the shop floor plus what's in the plant warehouse should cover all the orders to be transported in two weeks' time, plus the stock buffer. Let's look an example. We will assume that standard expected delivery time is 20 days, the production buffer is 15 days, transportation time is 10 days, the stock buffer is set at 100 units, and five orders, not yet shipped, appear in the logbook.

Order 21:    37 units. Ship to customer: 2 days from now.
Order 22:    19 units. Ship to customer: 5 days from now.
Order 23:    46 units. Ship to customer: 9 days from now.
Order 24:    35 units. Ship to customer: 12 days from now.
Order 25:    41 units. Ship to customer: 18 days from now.

Order 25 has 18 days left before we must transport, but the production buffer is only 15 days. This order is probably for a nearby client, which is why the transportation time is much shorter. Thus, the materials for Order 25 have not been released yet, and for this particular order we don't need the stock buffer. Orders 21 through 24 are already on the shop floor. We will assume that two older production orders are also on the shop floor. These orders actually represent replenishment back to the target level for earlier orders that were shipped from stock. These are:

Order 19:    30 units.
Order 20:    24 units.

Stock currently on hand is 46 units (that is, 100 minus Orders 19 and 20). Do we have a problem? Well, Order 21 is due to ship in two days. Even if production for that order isn't completed, there's enough stock to ship (37 required, 46 immediately available). However, if production for Order 21 or Order 22 fails to arrive within five days, we have a real problem. Order 22 requires 19 units, and we only have 9 remaining after shipping Order 21. Do we need to expedite? And should we expedite Order 21 or Order 19? How do we decide? In other words, how should we define a *red-zone* situation that would make production Order 21 (or Orders 19 or 20) urgent?

For MTO customers, the production time buffer that would have protected the transport time is 15 days. Therefore, we need a red-line time of 5 days. We expect to have the entire quantity required for the shipping during the next 5 days already at the warehouse. In this particular example, we need to ship 56 units (Orders 21 and 22) within the next five days. But we have only 46 units in stock. Consequently, we need to expedite Order 19. When this order is completed, we will have enough stock (46+30=76) to protect shipments for the next five days. So only Order 19 is a *red* order. At the moment, there's no need to expedite Order 20 or order 21.

Therefore, to determine a red-zone situation that requires expediting, compare the estimated time buffer required with the amount of stock on hand. When on-hand stock is sufficient for all shipments due within two-thirds of the designated time buffer, we're in the *green* zone. The proportion of total time the plant warehouse spends in *red* or *green* zones suggests when the stock part of the buffer should be adjusted upward or downward. Reality often squeezes the time buffer so much that introducing the stock buffer into the equation offers relief.

Managing two kinds of buffer—both stock and time— is complicated. It's worth considering only if the change required in the target inventory level results in dramatically less inventory to maintain, without compromising full availability. Unless the two-buffer approach described above results in at least halving the total inventory required, we recommend employing the simpler (and more straightforward) scheme of MTA. Additionally, we must consider whether it's possible to maintain a production time buffer that satisfies the market's delivery expectations.

In some situations, transportation time may be long compared with production time. In such cases, two time buffers may be used with an intermediate due date for the first buffer. Doing so protects the minimum transportation time from consumption by excessive production time.

## Simple Make-to-Stock: No Commitment, No Priority

Sometimes we may find that we have capacity temporarily available and enough market demand to buy what we can produce once products are available for sale. The big difference between this situation and MTA is that there's no commitment whatsoever. It also means what is produced is probably not very profitable.

Products that are really financially rewarding are normally sold with some kind of commitment, either for timely (preferably fast) delivery or with guaranteed availability.

Promising a commitment to the market, and obtaining a premium value for such a promise, might seem to be disadvantageous. It certainly mandates less-than-perfect utilization even from a CCR. But demand fluctuation and the inevitability of Murphy's law in production operations makes it crucial to maintain protective capacity. However, limiting sales because of a CCR—even when that CCR is not fully utilized—could compromise meeting commitments to the market in general. Making items to stock strictly to realize better CCR utilization, without any real commitment to the market, may relieve some of management's frustration, but it might add only a little to the bottom line.

One possible alternative is to create a clearly defined category of "no priority" items. The opportunity to further exploit a CCR's capacity depends on having demand without any priority attached to it. This ensures enough protective capacity to meet all the "real" commitments. We then produce the "no priority" items to stock only when there is absolutely nothing else to do.

Buffers must be very carefully monitored when producing "no priority" items to avoid threatening any commitment-based production. The three basic buffer management colors (red, yellow, and green) might even evolve into four, with the addition of *light blue* (no priority at all). Green definitely would have priority over light blue. For this to work, several rules would have to be established:

1. The items to be included in the "no priority, no commitment" group must be clearly defined. The two most important criteria are:
    a. The items can be easily sold from stock when they're available for an acceptable price, and customers would have no difficulty obtaining them elsewhere if we don't have them in stock.
    b. Production can easily begin processing these items whenever capacity is available, and stop when it's necessary to revert to "commitment" orders. Raw materials must be standard and capable of stocking at the CCR for its use whenever time allows.
2. Production must have an effective procedure to identify the situations when these items can be produced. To prevent having a warehouse full of items that can't be sold, this procedure should also limit the quantity of any one item that production is permitted to do. A target level for such items should be established, not for forcing replenishment but to ensure that production ceases before too much stock of that item accumulates.
3. Sales people must clearly understand the difference between stocks maintained for MTA and the stocks for immediate sale "whenever the opportunity arises." Sales should *never* sell MTA items as if they are no priority/no commitment ones.

The idea of augmenting actual make-to-stock with make-to-availability (or even with make-to-order) has led to rise of the term *market buffer* at some TOC conferences. But the term market buffer doesn't apply in this case. The intention is to exploit the capacity of the CCR while still preserving ample protective capacity. Having no-priority/no-commitment items produced doesn't protect the market, so no such market buffer is involved here—only an exploitation scheme.

## Reference

Goldratt, Eliyahu M. Personal conversation with Eli Schragenehim in Israel, 2003.

# Chapter 8

# The Theory Of Constraints Approach to Distribution

## Contents

Manufacturing a finished product is only half the battle. The other half is getting it to an end user in a timely way. As we've previously indicated, a complete supply chain includes raw material suppliers, manufacturing, and distribution to customers. In the preceding chapters, we've addressed the manufacturing part of the chain and described the process of tying the "rope" from customer demand to material release. However, this is only part of the solution. In this chapter, we will see how a Simplified Drum–Buffer–Rope (S-DBR) production operation integrates seamlessly with the distribution end of the supply chain.

# Distribution Component of Supply Chains

Producing products alone does not assure a sale, even if such products are actually in demand. Somehow these products must reach the end user, otherwise they will be stuck in a warehouse somewhere. Without sales to consumers to draw down finished inventories, future production of these products would ultimately end.

This is where distribution networks come to the rescue of the producers, at least in the case of consumer products. The task is to bring the products close enough to the consumer so they can be readily purchased. The challenge is to do so without clogging the entire supply chain with too much inventory or running short of finished products sought by customers. The solution to this challenge is not intuitively obvious to most companies for which supply chains are lifelines.

Completing sales of finished products involves three distinct tasks:

- Transportation
- Storing (including point-of-sale display)
- Selling

Most of the time, transport and storing of products are done several times on the often long path from a production floor to a retail shop. These move-and-store operations are usually affected by distribution networks. Often such distributors also operate the stores where goods are sold.*

---

* Well, some of those goods are eventually sold. Some, maybe not.

## Comparing Production and Distribution

Like production, distribution is a type of operational environment. Although it doesn't change anything in the product, distribution *is* concerned with utilization of resources, primarily transportation, space, and ultimately cash. Distribution networks can be quite complex, especially when limited shelf life is a factor, and distribution is certainly subject to Murphy's law.

What makes production complicated is the dependency among resources. The more dependencies, the greater the probability that production flow will break down or be interrupted in some way. This is one reason why the inventory between operations (work-in-process (WIP)) is sometimes perceived as "protection" against Murphy—it effectively insulates one operation from the special-cause variation that might affect its input. Consequently, an assembly operation appears to be a sensitive operation because it depends on the availability of several inputs simultaneously to do its job. Setup time, of course, contributes to the perception of complexity. And the pressure to achieve efficiency—maximizing every bit of productive capacity— can cause frequent conflict with sales functions, whose primary interest is to provide everything for which a customer asks.

An examination of distribution operations reveals intricacy that may look different, but is actually quite similar to production operations. The possibility of experiencing a stock-out depends on a number of factors. The quality of the supply might be one. Another is that transporters usually aggregate many items together on a single vehicle in the interest of efficiency. (This is precisely the idea underlying batching in production.) Intermediate storage space and the shelves at a point-of-sale are usually kept fairly full as a hedge against unanticipated high demand.

A common assumption about complexity is that the number of variables contributes to it in a significant way. In production, for example, every end-item is linked with many variables detailing how that item is produced: materials, resources, processing times, setup times, manpower and the skills they need to master, the routing sequence, quality checks, etc.; all of these are merely part of the huge database of "key" production data.

The comparable information factor in distribution is the number of items being managed and their locations. A typical distribution network may hold many thousands of items at every location at any given time. Every item and location is the subject of a critical decision: how much to store? The sheer number of these decisions is what makes managing distribution complex and often results in confusion.

Decisions on how much to store from item X at location Z are not trivial. They're really about whether to hold more or less. If you hold more, then the investment in inventory for that particular item is large, and the return on that investment is not always certain. Moreover, when space is limited, a decision to maintain more of one item could be at the expense of not maintaining some of another item.

Also cash limitations might mean that maintaining more of one item limits the numbers of other inventory items.

Holding less of a given item incurs a higher risk of running short. Because the whole objective of distribution is to try to ensure availability of what is demanded, any shortage certainly reflects a failure to reach that objective. Yet shortages *do* happen, and this in turn increases the pressure to raise inventory levels so as to avoid recurrences. Of course, financial investment then rises and the other adverse impacts of having too much inventory also increase until an opposing pressure to *reduce* inventory develops. Normally two such opposing pressures force the creation of a compromise. Inevitably, however, the compromise is short-lived. In the face of volatile market demand, these decisions require almost continual reassessment. The total number of items presents a unique problem in distribution.

The central factor in deciding specific stock levels at specific locations is the uncertainty associated with variability—in demand, production, and replenishment time. While in making to order uncertainty is a factor to be considered, in distribution it is *the* key issue.

## Measuring Distribution Network Performance

The most striking problem in distribution networks is how to exploit the available cash. Where the objective is to provide acceptable availability of a very large number of items, spare cash can be used either to increase stock or to increase the variety of items available for potential customers. The former option can ensure fewer shortages. The latter offers potentially greater customer appeal.

If cash is the active capacity-constrained resource (CCR), better exploitation of the constraint translates to higher inventory turns. Inventory turns is a financial measurement defined as *the annual cost of goods sold divided by the average aggregate value of inventory for that year* [Russell and Taylor, 2009] (Figure 8.1).

The reciprocal value of this number represents the average length of time an item would remain stored until sold. For instance, let's assume that a certain store turns its inventory four times a year. This means that a piece of inventory averages staying on the shelf for three months before it is sold.

$$T_i = \frac{C_{gs}}{V_a}$$

$T_i$ = Inventory turns
$C_{gs}$ = Cost of goods sold
$V_a$ = Aggregate annual inventory value

**Figure 8.1  Inventory turns.**

From a financial point of view, it might make more sense to measure annual sales divided by the average inventory value because this would be a better measure of the time required to return for the dollar spent buying the inventory. The two numbers are definitely not equal and they highlight a somewhat different aspect.

Suppose the average selling price in the distribution network is twice the cost of purchasing the item. If we continue the example of four regular inventory turns, the annual sales divided by the average inventory would be eight, meaning that any dollar spent on inventory is amortized after 1½ months. Naturally, this is a prime company metric. What's the value of doubling the inventory turns? Assuming you are still holding the same amount of inventory (i.e., you have the cash and the space for it), it means doubling the sales, which is equivalent to doubling the throughput. The net impact on the profit is even more striking.

## *Managing Inventory Turns*

What keeps us from turning a knob to set (or change) the level of inventory turns? Could it be the fact that we do not know with confidence what will sell well in the months ahead? Obviously, with a crystal ball to predict the future most decision making would be so much easier. Managing distribution networks would be simply technical, to the point that computers could do it. However, because we are unable to predict the future accurately, a lot of effort goes into forecasting in the hope that it might tell us something about the future that could be useful in making such decisions.

But is difficulty in predicting demand the only missing information for effective decision making? As we've already seen in Chapter 6, two diverse and somewhat randomly behaving factors affect inventory: demand and supply (Figure 8.2).

System behavior over time in both demand and supply is far from being comfortably predictable. Distributors' decisions create the causal connection between the two: when demand is high, the distributor asks for greater (and faster) supply. When demand is low, the distributor wants to reduce the input of supply. Eventually the mismatch between demand and supply is reflected in the accumulation of some inventory and shortages in other inventory.

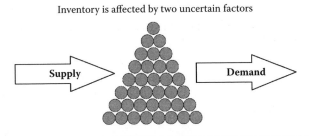

Inventory is affected by two uncertain factors

**Figure 8.2   Inventory links supply and demand through time.**

Unlike the situation in make-to-availability, the possible lack of production capacity is not usually a significant contributor, unless the distribution company owns the production process as well. Frequently, distribution companies purchase from other suppliers—most of them not dedicated suppliers—and thus have no control (nor any real need for it) over the supplier's capacity management.

Large numbers of end items and numerous dispersed locations force most distribution networks to emphasize less the speed and frequency of the supply. Consideration of resupply time in determining the target stock level happens only at a gross level. And if the supplier is located on another continent, the stock must be higher. The question is: by how much?

An even more challenging question is how to determine the stock to maintain for such a large number of end items and locations? Moreover, how should those decisions change as market demand keeps changing?

Behind these troubling questions is an underlying problem: cash is limited. Most of the time cash is *the* constraint. Space may also be limited. Each decision might be focused on one item at a specific location, but the accumulation of all decisions can easily clash with either cash or space limitations. If this happens, what can be done about it?

## The Role of Forecasting

Forecasting plays a very important role in the traditional management of distribution networks. A forecast helps predict how many items will be sold in future planning periods. A sales forecast is also an input for determining the production target levels. It's relatively common (and easy) to let a computer system dictate a sweeping decision to hold, for example, two months of inventory for all end items in a large category of products—assuming that all items have comparable replenishment times.*

Unfortunately, this assumption frequently is not valid, especially when different suppliers are involved. The inevitable result is that all stores experience frequent shortages, while having too many of the other items that don't move as quickly. Haven't we all experienced this phenomenon personally? How many times have you gone to a store with a list of items to buy and left without several of the items on your list? At one time or another, each of us has been faced with a decision either to accept a less-than-desirable replacement or to go somewhere else, without any assurance that the missing items will be there.

At the same time and location, a clearance might be under way, in which the company puts "on sale" all the items it considers in excess, and these items typically occupy much of the display space. This is just a common indicator of the inventory imbalance in the distribution network—too many of the wrong items and not enough of the right ones.

---

* The authors have observed this in more than one industry.

## Excess Inventory for Distributors

Excess work-in-process in production is devastating because it overloads nonconstraints and blurs the priorities, but is it as devastating in distribution? After all, most of what is not sold at the regular price will eventually be sold at a reduced price. At the end of the day, the majority of the items in a supply chain are sold, most of them with a positive margin. So what's the problem? The real problem is that *too much inventory of a specific item definitely adversely impacts sales of some other item*—an item that very likely could be sold at a higher price.

Let's examine a proposed cause-and-effect relationship (Figure 8.3). Suppose the local manager of a regional warehouse finds that the warehouse is holding too much stock of Item X. They're too slow-moving. Each time the manager walks by the small mountain of Item X stock, he feels uneasy. Eventually, he feels compelled to do something about it.

The most obvious thing to do is to push more of Item X to the retailers. All salespeople are directed to persuade retailers to take more of Item X. First, the sales force tries to convince retailers how great Item X is. At some point, they might use the power most influential suppliers have on their customers. When the additional sales of Item X still don't move fast enough, the local manager might offer an attractive price discount to the retailers, based on volume and lasting only for a short time. This usually does the trick—retailers take the bait, and the mountain of Item X gets smaller, perhaps even vanishing completely.

So what's the problem? There is no real loss—or so it seems.

Let's look a little farther down the supply chain and see what happens to the retailer, who bought a substantial quantity of Item X. The retailer also wants to move Item X. After all, X was a slow selling item in the first place. So the retailer dedicates a lot of space to displaying Item X, and, of course, passes on the special discount to customers. And it works. All of the X items are gone. Of course, this sale of a slow item doesn't continue forever. It was specifically targeted to move Item X out of the supply chain while still making a profit, nothing more.

What is wrong with this commonly used scheme? *The unavoidable result is that sales of other items—possibly a substantial number—have been lost because of the extraordinary push to sell Item X.*

Let's suppose that Items Y and Z are in the same broad family of products as X. But both Y and Z are in much higher demand than X. The problem is that the original forecast predicted better sales for X than actually materialized—at least until the local manager of the regional warehouse decided to give X a push. How are the sales of Y and Z impacted by the push to get rid of X?

If no exceptional push had been given to X, sales for both Y and Z would be higher than what they are likely to achieve when X is emphasized.* The efforts of

---

* This is based on the quite reasonable assumption that consumer spending power has limits, and purchase of one product is likely to be at the expense of a comparable product (same product family).

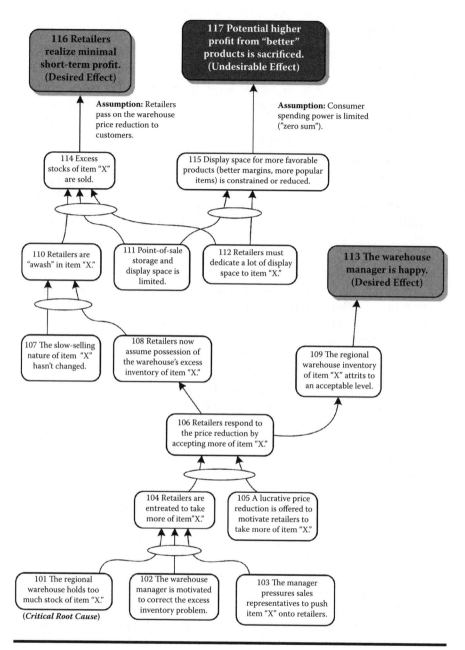

**Figure 8.3  Profit impact of inventory imbalances.**

the sales force in pushing X usually come at the expense of selling or servicing the other products, notably Y and Z. While Y and Z sell well, retailers have limited space and cash, so when they're faced with pressure to buy more of X, this unanticipated purchase comes at the expense of the sales of other items (at the very least, less effort to sell). Retailers normally aren't favorably disposed to simply increase their total purchasing. Similarly, when the retailers dedicate more space to X, it is often at the expense of displaying Y and Z. Of course, cheaper products "steal" sales from premium-priced items.

Thus, while experiencing shortages is obviously bad for business, so, too, is excess inventory of a slow selling item.

## Stock-outs: Damages and Opportunity

It's important to consider the impact of a stock-out in retail. A manufacturing organization that commits full availability of a certain mix of items to specific clients could gain a decisive competitive edge for doing so. For some clients, such a company could offer vendor-managed inventory (VMI) to ensure availability at the client site, not just at the plant warehouse. This kind of commitment is a very serious matter because any failure to fulfill such a promise could have devastating ramifications.

Managing for availability within the distribution channel does not seem to be as strict a commitment to the customers as it is between a manufacturer and its business clients. Offering a large variety of end items means that a certain amount of shortages are to be expected. This is especially true of stores where clients comprise a variety of small potential customers. Too many stores assume that if, in the face of a stock-out, they offer alternative items customers won't be too disappointed. However, at some point, clients are disappointed to the extent of preferring to shop in another store—it's not always assured that they will accept the alternative offered.

It's clear to everyone in distribution that stock-outs are generally bad for business, even though there may be some cases where no real harm is done. For instance, when the client doesn't perceive much difference between an out-of-stock item and an alternative item that's available, the damage, if any, is perceived to be negligible.

It should be noted that every shortage causes statistical damage to both the distributor and to the producer, and that the damages to each are not the same. If a customer finds an acceptable alternative product at the point-of-sale, most of the damage accrues to the producer. But if the customer is intent on searching for his beloved brand at another store, most of the damage accrues to the rest of the supply chain. Statistically speaking, however, both are harmed by a shortage.

In some markets, experiencing a stock-out is interpreted as almost a good thing. The demand for items in fashion, for example, fluctuates even more than with items that seem more regular. When, for whatever reason, a certain product becomes a real star, it vanishes quickly from shelves. If you happen to look for such a popular product, you might experience a typical response from the salesperson: "Oh, we're sold out of this product." There is often a kind of pride in saying that the item is "sold out." The implication is: "You should be quicker to come in next time." But most of the time, a complete sell-out of an especially popular item isn't something that marketing people planned to happen. Thus, in a chain's warehouses, a sold-out item represents a mistake.

Distribution in the fashion industry is unique because it divides the year into seasons, and each season has its own completely new products. The most common practice is to have all the products for next season available at point-of-sale at the very beginning of the season—sometimes even before. As some new designs become extraordinarily popular, such "stars" sell out relatively quickly, while most other new items might experience only mediocre demand and will end up in season closeout sales.

This example is especially interesting because of the real pain of failing to be able to capitalize on so much of the market demand for products at a premium price. It clearly refutes the idea that being "sold out" is a good thing.

What about stock-outs that customers are not aware of? The supply chain managers may be aware of some of these. The plan may have called for a particular item to be in stock, but for some reason it turns up short at a particular location. If customers don't actively search for that particular item, such a shortage might not seem to cause any damage. But this is not really true.

In spontaneous sales, disappointment isn't really relevant. Customers aren't likely to be troubled by not finding a certain item in the store if they weren't specifically searching for it, unless they find something they like but not in the correct size. Even then a stock-out can mean statistically losing sales. Some products are "targets of opportunity" for customers who are coming into the store without a specific idea in mind about what they are looking for. They are only browsing, but the availability of products to buy can convey to the customer an attractive variety that includes items they might buy.

It's not easy to predict which items are going to be stars and which will be dogs.* It naturally follows that star items are likely to represent a small percentage of the entire variety carried, and they are likely to be exhausted fairly quickly. Even if customers aren't aware of the absence of the star, they might still find the variety lacking because of it.

The lesson is that maintaining better availability while realizing higher inventory turns can do wonders for a business. We will try to quantify the potential impact on the bottom line, but first let's consider when it might be possible to do both of these things.

---

* The terms *stars* and *dogs* come from the Boston Consulting Group's widely known classification of product portfolios into four categories—stars, cash cows, question marks, and dogs—based on their profit-generating potential.

## Availability and Increased Inventory Turns: Some Insights

Managing stocks in distribution is generally similar to managing stocks in production. One considerable difference is that the sheer quantity of items involved is much larger in distribution than in manufacturing. Another difference is that most of the time the suppliers for a large distribution network are not part of the same organization, thus each has less direct impact on the supply. That said, however, in many cases the power of the distribution network on its suppliers is considerable. Actually in many cases the impact of the distribution on the producers may be even larger than when the distribution and production are part of the same company.

The impact of suppliers is important. Focusing on the supply is the first insight Theory of Constraints (TOC) has to offer. This is not a huge revelation. Everyone in distribution can see how improving the supply response time would positively impact both availability and inventory turns. Moreover, all large chains put a lot of pressure on suppliers, demanding short lead times and reduced prices. But pressuring for a low price and a very short response time simultaneously don't always go together effectively. The linkage between immediate demand and supply is always there.

Insight 1: Shorten the supply response time as much as possible.

Insight 2: Capitalize on a shorter, more reliable response time to really lower the inventory level and to respond quickly to any "almost short" signals received.

Insight 3: Stabilize the inventory levels per item/location. Do not continually change them based on small fluctuations in the forecast.

These are by no means all the insights required to achieve the ambitious target of improving availability while increasing inventory turns. But for now, they're sufficient for us to pause and analyze what these insights contribute to distribution management.

Insights 2 and 3 suggest linking the inventory in the system to the response time, but they discourage changing inventory targets too frequently. A way to accomplish this linkage is to fix an inventory target level *based on the replenishment time* and replenish it when it goes below the target. That target level is updated only when a clear, consistent signal is observed that the inventory is too high or not high enough.

Now, it might seem that this approach simply describes a common way to manage stock known as the min–max method, or reorder point. Both of these ways define a reorder point. When the inventory drops below that reorder point, a resupply order is issued to the supplier, requesting replenishing the stock by either a fixed batch (according to the reorder point method) or to a predetermined maximum level (the min–max method). So what is new in our approach?

Apart from determining target levels based on replenishment time, what does Insight 1 tell us? *Very simply, if we find that some component of the replenishment time has nothing to do with supplying the replenishment order as fast as possible, we should simply eliminate this unnecessary delay.* In the max–min and the order-point methods, there is such a delay. This is the requirement to wait for the consumption of a minimum batch before a new supply order is issued.

Figure 8.4 shows the changing condition of actual on-hand inventory at a typical distribution location that uses a min–max rule for reordering. As sales occur over time, on-hand inventory decreases. When the decreasing inventory penetrates the reorder point level (minimum level), a purchase order for a predefined batch is issued. This may be the number of units needed to replenish inventory to the maximum level. However, even though the purchased order has been issued, it still requires time for the order to be produced and delivered. During that period, inventory consumption presumably continues at some rate. When the order is ultimately received, it increases actual inventory level again, but not quite up to the maximum level. The total replenishment time is built from (1) order-time—the time from the perceived need to reorder until the purchasing order is issued—and (2) the actual resupply time until the order arrives.*

Before we try to improve resupply time, let's verify whether the order time is really needed. The damage from having that time as significant part of the replenishment time should be obvious—a significant delay in the replenishment leads to maintaining too much stock and being slow to respond to a sudden surge of demand. If a supplier requires three weeks to deliver an order, and if we batch-process our reorders weekly, we could be increasing the replenishment time by as much as 33 percent. How much more inventory does this drive us to maintain to protect against stock-outs for that additional time? In most cases, the time delays associated with both batching and processing the purchasing order apply. First, new purchasing orders are usually done once a week. Second, because the min–max is still in place, most items aren't ordered weekly, but usually once during several weeks. These practices expand the actual replenishment time considerably, and the resulting damage can be very serious.

## Batch Size versus Buffer

Batching is typical to many types of operations. Managers with an obsession for efficiency consider increasing batch size quite natural—their objective is to save costs. There are two penalties for doing this. First, the effect of batching replenishment orders is to increase replenishment time unnecessarily. Second, and more importantly, with a min–max policy most of the inventory maintained (the unshaded area under the curve in Figure 8.4) incurs a substantial investment cost,

---

* Companies typically strive to make all parts of their operations "efficient," including purchasing. Consequently, a purchasing department may batch-process reorders, for example, on a weekly basis. This can incur an administrative delay of days or weeks, depending on the individual company's reordering policies.

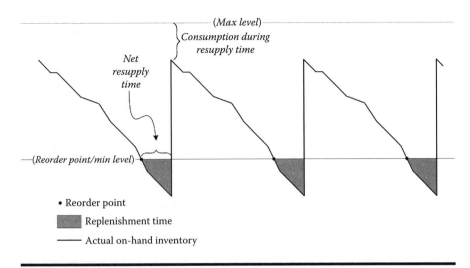

**Figure 8.4  Min–max order time.**

but adds no real value. As we have seen, the additional inventory pressures local managers to push the sales of particular items at the expense of more lucrative items. Notwithstanding the detriment to the company as a whole, the purchasing department is definitely more efficient.

TOC (and Lean, as well) ask the obvious question: what cost is really saved by batching? If all it represents is less work realized from people who aren't currently fully loaded, then let's stop batching. There's no justifiable reason to do it. Moreover, compressing replenishment time—taking out time delays that aren't really needed—can actually make the larger distribution system much more responsive to changes in market preferences.

This is not intended to imply that batching, per se, is necessarily a bad thing. Some batching will inevitably be required. Batching normally is motivated by at least one of two parameters: *quantity* and *time*. A company may do quantity batching to achieve efficiency in transportation. An ocean container or a tractor-trailer would not be dispatched to ship a single 1-cubic-foot item. In this case, the batching refers to a total volume, composed of several items combined into the one batch dictated by the size of the container. More likely, resupply items would be queued until they reached a certain cost-effective volume (or weight) for shipment. A company might also do batching based on time (frequency). This is purely a question of administrative efficiency. In these days of computerized inventory management and order-entry systems, it's difficult to justify batching the entry of dozens of orders once a week rather than entering the reorder data as it comes in. There is no cost saving, but there is a potentially large (though usually unrecognized) inventory cost incursion in doing so.

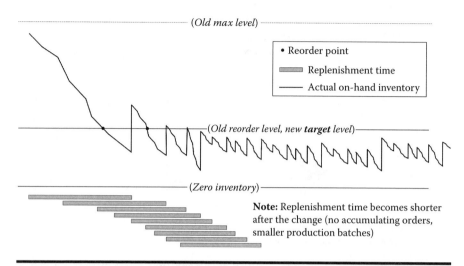

**Figure 8.5   Daily reorder submissions.**

Suppose it's practical to make the order time very short—one day, for example. What would the impact be on stock in the system? Each day a purchase order would be issued for replenishing exactly what was consumed the preceding day. The order would normally arrive after the actual resupply time (from the time the purchase order is issued to delivery of the order). But if we issue a purchase order each day, and the supplier responds to it without adding batching of his own*, then, on an average day, a shipment will arrive containing the items and quantities equivalent to what was sold one resupply period ago. The actual level of on-hand inventory with such frequent shipments is fairly stable. Moreover, the total stock the company has committed to—including on-hand stock and all orders in process but not yet received—becomes very stable near the target level.

Figure 8.5 illustrates the inventory situation when batching is abolished. At first, when the change is introduced, the first shipment shows up only after the calculated resupply time, but after that, every day brings in another shipment.

Insight 3 discourages changing the target inventory level too often. But this insight raises questions. What does "too often" mean? And how do we know *when* the target level should be modified?

The target level must cover for possible demand within the replenishment time. Replenishment time is defined as *the average time from a sale until the piece replacing that sale has arrived.* The "possible demand" allows for a *liberal estimate of the most optimistic sales that might occur within the replenishment time period.* The target level should also protect against certain potential delays in the replenishment time.

---

* And assuming no delays in the traffic from the supplier to the distribution location.

Don't confuse this requirement for the target level to cover all the sales within replenishment time with the time between shipments. As Figure 8.5 illustrates, the replenishment time (represented by the stacked gray bars) is longer than the time between shipments (daily). Higher shipment frequency dampens the impact of fluctuations in demand. However, it doesn't protect the availability from the impact of unanticipated increasing sales.

Suppose the average sales per day are normally 100, but today is 150. The on-hand stock will be down by 150, but the next shipment—and succeeding ones—would be around 100 until the shipment specifically initiated by today's sales of 150 arrives. As long as the sales tomorrow and for succeeding days revert back to around 100, the extra 50 shouldn't pose much difficulty during the replenishment time of the order for 150. However, if consumer demand remains near 150 for an extended period, an upward adjustment in the target level is clearly indicated, but not until the increase shows signs of being more than just a temporary aberration.

Remember that the buffer, referred to here as the inventory target level, is a liberal estimate of what is required. "Liberal estimate" means that the buffer is able to cover for a significant number of changes in demand. Consequently, *the buffer should not be adjusted to the forecast,* unless the change in forecast is truly substantial, such as occurs during a very strong seasonal peak. Let's emphasize this point: *frequent forecasting introduces changes that could be just a reflection of the regular "noise" in the system.* Adjusting the target level to natural forecasting fluctuations increases the total noise in the system. W. Edwards Deming referred to this as "tampering" with the system—(inducing variation unnecessarily, through ignorance, in a system that was actually within normal statistical control limits) (Deming, 1993).

## Increasing the Buffer

It's clear that sometimes the buffer really needs an adjustment, meaning that the inventory target level should be aligned with reality. This brings us to the next insight:

> Insight 4: The behavior of the on-hand stock reflects whether there is a need to adjust the size of the buffer.

This insight directs our attention toward on-hand stock. Instead of reforecasting based on an optimistic demand and verifying whether replenishment time has changed, it's sufficient to monitor the on-hand stock—the part that provides the active protection of availability. This provides us a very clear focus on the combination of demand and supply—exactly what we need to assess whether the buffer is "about right" or not.

How does on-hand stock vary? Consumption through time reduces its level, while supply increases it. Now if the target level is too small, either because the

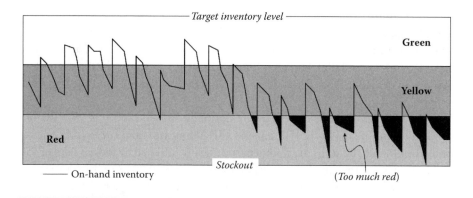

**Figure. 8.6  Long and deep "red" penetrations.**

demand is much larger than expected or the supply time is longer than normal, then for some period of time the on-hand level will remain relatively low.

In production buffer management, we refer to the last one-third of the buffer as the "red zone." The same is true for inventory buffer management. So if the buffer is too small, we would expect to see on-hand stock remain in the red zone for a significant amount of time. It's "too much red" when the on-hand inventory remains too long, and too deep, below the threshold between red and yellow.

Figure 8.6 illustrates the situation. Consumption is generally continuous (low consumption during short time periods), but shipments are made in a specified batch quantity (such as daily sales). We can see that the graph shows a downward trend, which is visible only over a longer period of time because of the visual impact of the short-term (daily) inventory level fluctuations. The fluctuations are confined to a narrow range. Eventually, toward the right side of the graph, the penetrations into the red are deeper and longer than before. This suggests why we refer to that situation as "too much red." But what should be the clear rule? We certainly need one and we should follow it rigorously.

Goldratt suggests the following approach for determining when the red penetrations are "too much": *if the total number of red-zone penetrations (by units) during the calculated replenishment time exceeds the value of the red zone, the target level (including the red-zone, maintaining its 33 percent ratio of the target level) should be increased.*

Let's use an example to explain this (see Figure 8.7). We will assume our replenishment time is 15 days (three work weeks). During a rolling 15-day window, we will analyze each order that drives the inventory level into the red zone and count the number of units that qualify as red. Now, at short intervals (possibly daily), we will also receive stock replenishments that may increase the inventory out of the red zone, or at least reduce the number of units "in the red." The inventory during that 15-day period will rise and fall. But we will continue to accumulate a daily tally of units in the red. At the end of 15 days, we will compare that total with the number

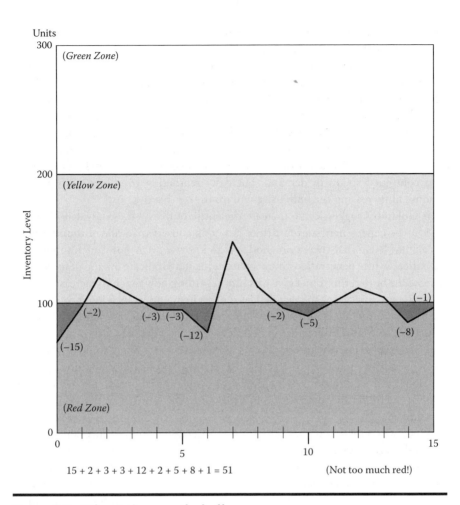

**Figure. 8.7   When to increase the buffer.**

of units that make up the red zone. If the red zone value is exceeded by the number of units in the tally, the entire buffer should be increased by 33 percent (equally distributed among the green, yellow, and red zones).*

The rationale for this is that combining the depth and length of the red zone penetration provides a more realistic measure of urgency. Establishing the buffer-increase signal at the time when the 15-day total of daily penetrations equal the red-zone size is, of course, somewhat arbitrary. But it also means that whenever a stock-out occurs, it's already "too much red."

We use the replenishment time—the time from reorder to receipt of the items required to bring the inventory up above the level of the red zone because

---

* This rule of thumb is based on a conversation between Eli Schragenheim and E.M. Goldratt in 2003.

it represents the critical period during which the buffer must be large enough to absorb routine fluctuations in demand.

When the accumulated buffer penetration occurs much faster than the replenishment time, we can initiate the buffer increase immediately. For example, if the red zone is 100 units and the total penetrations reach that figure within five days, we don't need to wait for the full 15-day replenishment time to expire before deciding to increase the buffer size. It should be recognized, of course, that such an increase in the buffer level would require one replenishment period to take effect. During that replenishment time, stocks are at somewhat greater risk of depletion from continued spikes in demand, but daily reordering combined with a rapid response in increasing the buffer size will minimize this risk.

Is Goldratt's suggestion to increase the entire buffer by the equivalent of a full zone (33 percent) a hard-and-fast rule? No, it's a conservative rule of thumb. First, let's acknowledge that there's no good way to know *a priori* how much to increase the buffer. While penetrations into the red zone may indicate an apparent need to increase the buffer, they don't give us a hint regarding how much to increase it. This is because the forecast at the detailed level of an item and its location is too inaccurate to provide any useful information for decision making. On top of that, replenishment time is also imprecise. Sometimes it may be relatively fast, other times it might take substantially longer, and the variability is not likely to be predictable. Consequently, we must ensure that the buffer is not too small, or not too large, but just right.* All we can do is make the best possible initial guess, then continually monitor the buffer for some period to determine whether that initial value should be modified.

Once we've increased the buffer, some patience is required in evaluating the new buffer size. Absent a series of drastic spikes or dips in demand, we should avoid the temptation to evaluate and adjust the size immediately following an increase. Because of the time lag between increasing the buffer and a resulting increase in on-hand inventory, such adjustment constitutes tampering by Deming's definition of the term. When the buffer is increased, the immediate resulting action is the issuing of a new order for the net size of the increase. A practical rule is to defer checking again only after a complete replenishment cycle following the increase of the buffer.

## Decreasing the Buffer

Thus far we've defined the indications that could lead to increasing the target inventory level (buffer). There is also a possibility that the target level might require reducing as well. Excessive inventory can be detrimental to overall system financial performance, too. A decision to decrease the buffer should also be based on the indications of the on-hand stock.

---

* Just like Goldilocks' porridge!

Suppose our inventory target level is too high. What unavoidable consequences would we expect to see in on-hand stock levels? Because sales would be low compared with the stock in the system (as defined by the target level), we could expect on-hand stocks to remain relatively high. Certainly if stock levels reside in the green zone for extended periods of time, it's a signal that the protection provided by the stock (combined with the demand and the suppliers' response) is higher than required.

Similar to the rule of thumb for increasing inventory buffers, Goldratt suggests *reducing* the target level—by a full zone, 33 percent—only when the time on-hand stock remaining in the green zone equals the replenishment time.

At the time the decision to reduce the buffer is made, the normal situation is that the current stock, which may have been in the green zone before the reduction in the old buffer, is now likely to be *well above* the green. Before we start to measure the time on-hand stock is in the green (for monitoring purposes), we must allow it to draw down until it crosses into the top of the green zone again. During that attrition, no new replenishment orders should be issued because, according to the new definition of the target level, the stock in the system is above the target level. However, once the on-hand stock drops below the new target level the replenishment scheme is initiated again, including periodic checking for "too much green" to determine whether the target inventory level should be further reduced.

## Dynamic Buffer Management Rules

The process of adjusting the stock buffer is called *dynamic buffer management* (DBM). It's a key element in the TOC approach to managing a huge number of different buffers. (Each item/location defines a unique stock buffer.) In reality, the size and complexity of distribution networks force the need for automated management of changes in target inventory levels. The sheer number of the decisions involved means that a human responsible for even one category of items at a specific location would be too overwhelmed to exercise sound judgment every time one item is either "too red" or "too green." Instead, if a computerized system is programmed to follow the algorithm and make the changes, the responsible manager must only briefly review the list of changed buffers to see whether a few of the changes ought to be questioned. Deferring DBM to a computer seems prudent, perhaps even necessary, in such situations.

It's worth emphasizing this key difference between managing stocks in distribution and in production. In a manufacturing environment the number of items under management is far smaller. The possible negative consequences of a buffer change, even when the algorithm seems sound, are considerable. Remember, the risk of running out of capacity—a huge risk in production—doesn't really matter in distribution. The risk to a cash constraint is much lower because increasing stock buffers usually signifies that total sales are increasing. When this happens, the cash constraint is usually not so severe anymore. But if cash is close to being fully exhausted, the entire management team should be highly sensitive to approving anything requiring more cash, such as an increase in inventory buffer level. At such time DBM might be temporarily suspended, but under normal circumstances DBM should be automated.

Still, any automation producing significant financial impact should not be adopted casually. Computers have no intuition for identifying "suspicious" situations and reconsidering existing procedures afresh. Thus, DBM, more than in any other practice in constraint management techniques, demands careful judgment to control it carefully.

## Deactivating DBM

There are two general circumstances when we should not consider automatically programming buffer changes. The first occurs when we know something about future demand that is likely to be at odds with DBM, which is only backward looking. The second situation is the identification of a supplier problem. There are, however, other situations that don't fall conveniently under these two broader headings. Let's examine all three situations.

### Future Knowledge

Occasionally, we anticipate a demand surge that doesn't correspond to existing demand, for example, when we expect a seasonal change, or when we change our marketing to boost sales (promotion or a special advertising campaign). In these cases, we usually expect higher demand to precipitate, even though we are currently experiencing much lower demand. Before a high demand onset, DBM might actually tell us to *decrease* the buffer, when we believe that the buffer should actually be increased. At the same time, when we increase the buffer in anticipation of a surge in demand, we must ignore the "too much green" signals clamoring for a buffer reduction.

Similarly, when a decision is made to discontinue selling an item (perhaps to replace it with a newer model), the target inventory level should be manually reduced, in spite of what DBM might be telling us, to prevent further replenishment when sales are abruptly stopped.

Dealing with the future knowledge exception requires signaling DBM to refrain from changing the buffers and to specifically "flag" whether this temporary suspension of DBM is for increasing or decreasing the buffer, or for both. This intervention should be done on the stock-keeping unit (SKU) level.

### Blockages in Supply

Suppose a certain location experiences a stock-out of an item. Should the target inventory level be automatically increased? What if the source—the warehouse that is responsible to supply that item to that location— is out of stock as well? Perhaps the entire supply chain is "empty" of the item.*

The basic assumption in determining the target inventory level is that the upstream node that sends the goods to a specific location provides reliable availability. If this isn't the case, the whole approach to the solution must be different. If a stock-out at the sending node results from a one-time event, it's better to continue under the current assumptions and the established target level. In such a case, the problem at the source must be addressed, and not just the resulting impacts.

When the source of the supply trouble is the producing node, increasing the buffer might actually be detrimental, inducing "noise" into the system that only complicates management. If the cause was lack of capacity in production, increasing buffers would only make it worse (as we have discussed earlier in this book).

Automatic DBM should be configured to highlight certain blockages in the supply chain and, instead of acting automatically and increase the buffer, send a message to the inventory manager. We assume that those cases are not all that common, so the manager in charge can and should devote the time to evaluating these specific situations.

## Other Times When Users Should Override DBM

The first of these is consumption of stocks resulting from expiration. Some products have a fixed expiration date. When management of these items fails to identify these expiration dates early enough for orderly drawdown, stock could be scrapped in an unplanned move. Such almost instantaneous elimination of stock could create an immediate problem of availability. Dated items typically expire when demand is usually very slow, so deep penetration into the red caused by too many expired units is not usually a valid reason to increase these items' inventory buffer.

Another situation is a commitment to a valued customer to hold a certain amount of items, no matter how slow they are consumed. Spare parts are an example of such a requirement. More commonly, however, some items may be for display in a shop. In the first case, DBM should not have any influence on stock levels. The latter case should define display quantities as an absolutely minimum value, over and above the basic inventory buffer.†

For example, suppose the target level is 30 units, of which 5 are defined as "absolutely minimum quantity." The stock in the system should be maintained as 30-plus-5. If we have 15 units on hand (including the minimum 5), the buffer status should be 67 percent: (30 + 5) – 15/30. In this way, the extra five don't con-

---

* Think of iPods®, for example, when they first came out.

† Retailers never want to be forced into a situation in which they must sell their display models. This signifies that they are too close to stock-out.

fuse evaluation of the buffer status, yet the existence of the extra five units is not completely ignored.

Items having small margins should also be flagged for a possible buffer increase if there is concern about the stability of their demand. However, the software should not be allowed to do this automatically so as to avoid unnecessary investment in a buffer that may be too high.

## Managing Transportation

Buffer management's main objective in production is to provide the right priorities in guiding the operators to decide "what to do *now*"? Another objective is to gain insight from the analysis of red (late or almost late) orders. One purpose is to identify blockages in the flow by determining which work centers have held more red-zone orders than others and similarly remediating the causes for excessive red zone penetration. Yet another benefit of buffer management is to signal a general increase in pressure on the manufacturing floor by monitoring the number of red orders in comparison with the total number of open orders on the floor.

In distribution, however, the benefits of buffer management are provided mainly by DBM, as just described. Adjusting buffers is a common and necessary activity in distribution, given the huge quantity of items/locations and the frequent changes in market preferences.

Yet the basic operations of distribution also require the same priorities used in production buffer management. In using these priorities for distribution, especially for decisions about transportation, it's useful to look at two different aspects of the buffer:

1. The normal picture, which includes only stock on-hand
2. An additional picture, which includes stock already in transit as well as what's on hand

Here is a simple example to demonstrate these two aspects. Let's say that Product X has a target level of 100 units. At the moment, stock on hand is 40 units, and 30 units are already in transit on a truck that's supposed to arrive in eight hours. The basic buffer status would consider only the 40 units on hand, meaning a buffer penetration of 60 units out of 100. This is a buffer status of 60 percent (still yellow).* But considering the items in transit as well, there are the 40 on hand plus 30 units in the pipeline (meaning in transit). We call the combination of on-hand

---

* The term *buffer status* refers to the size of the "hole" or gap in the required buffer, not the amount of remaining inventory. "Inventory level" represents the converse—stock that is available. In the past, some people have considered buffer status to be the amount remaining (available). We prefer to use higher numbers to indicate a greater urgency.

stock and what's in transit the *virtual buffer.*\* In the example, the virtual buffer status, considering both parameters, is 30 percent: $(100 - 40 - 30) \times 100/100$. This means that considering both on-hand stock and that in the pipeline (to be available shortly), the inventory level is still in green (above 66⅔ percent).

Theoretically, every time a consumption of even one unit occurs from a stock of one item at any node in the distribution network, an action to replenish that item should be commenced immediately. Thus, considering both on-hand stock and in-transit stock, we should expect to experience a "zero penetration buffer status," connoting that the total of items in transit and on hand is equal to the desired target level. If this always has been the case, then there is no need for the second perspective. In reality, this is not only not the case, but the in-transit parameter is an important input for the manager in charge of transportation from the source. The on-hand stock perspective is critical to the manager at the destination. Hence, the option of looking at the same buffer from different views is referred to as a virtual buffer. It provides more focused support for managers within the distribution chain.

In reality, delays could occur in acting immediately to replenish the system's stock to the target level. *It is highly recommended that distribution networks not delay a new purchase order to their suppliers.* Suppliers might batch that single request with others, causing some degree of indeterminate delay. But from the supply chain point of view, daily reporting is critical in minimizing stock without adversely affecting availability.

Let's consider the movement of goods within the distribution network for a moment. The first action required to replenish one node in the chain from a warehouse is to collect all the items. Doing this frequently for *all* the items to be transported between the points sounds much easier than it really is. Current common practice transfers a limited number of different items, each in quite a large quantity of units.

Adopting a new habit—transferring a large number of different items, each in relatively small quantities—represents a significant change in operations. The most tedious part is collecting all the items in a central location in the warehouse where they will be loaded onto a truck or van. Standards of packing might change as well, to facilitate effective loading and unloading of small quantities from the truck. There's also one more operational change: unloading a very large number of different items, validating and accurately registering the quantities, then properly storing them. This requires good organization. While this new procedure for loading and unloading is not exactly a paradigm shift, it certainly requires careful planning.

There are, however, practical limits to how small reorders between a warehouse and a point-of-sale can be parsed. When a store sells just one item, it's not realistic to expect that a replacement would be put on a truck from the distribution warehouse

---

\* This term was introduced by Inherent Simplicity Ltd., in 2006.

immediately. Even moving such an item the next morning might be impractical. So we can't assume that the virtual buffer will always show a status of zero.

Consider this example. Suppose a certain regional warehouse supplying a store is about an eight-hour drive away. Normally, the warehouse uses vans to move goods to the store, and such a van usually delivers to several stores on each trip. Would you dispatch such a van for a 16-hour round trip just to deliver a very few items perhaps no more than $100 in total value? What if the cargo value is over $10,000, but the van is still 75 percent empty—would you let the van go? What other information would you need to deliberate on such a decision?

What we are describing here is a batching dilemma: should we dispense with batching completely in order to respond as quickly as possible or should this inclination be sometimes qualified by common sense?

Transportation batching has a peculiar characteristic of being a kind of "global" batch. The batch is not usually for a specific item, but is likely to comprise all items that need to be transported in the same general direction. This batching dilemma is relevant when the actual cost to the company is not per piece, as it is in public transportation, but when the vehicle and the manpower are owned and operated by the company or contracted from a third party. The cost in these cases has little or nothing to do with the size of the cargo. It's based on vehicle plus driver(s) for a certain number of hours or miles.

TOC doesn't primarily emphasize cost reduction, but it can't ignore wasted cost either.* When TOC suggests replenishing as quickly and frequently as possible, some kind of balance is necessary. It was never intended to prompt irrational, possibly very expensive, action. Common sense is absolutely required.

What would be a commonsense approach to transporting items from one node in the chain to the next one? Two parameters must be considered:

1. How fully loaded is the transport vehicle when all required items are on board? Certainly, if the vehicle is 90 percent full, it should be dispatched. This is a typical reflection of resource utilization in production: load versus capacity. But in transportation, we have truly variable costs associated with every trip.

2. Are any of the items truly required *now*? This is a key question when the vehicle is relatively empty. If all items scheduled for transport are considered "in the green" when you review their virtual buffer status (meaning there are already sufficient items in the pipeline), it makes sense to delay the transport by at least a day. An "in the green" virtual buffer should be the prevailing situation: the combination of on-hand stock (at the downstream end) plus whatever is already in transportation pipeline should normally be in the green—and not too far below the established target inventory level.

---

* This is particularly true for transportation if fuel costs rise steeply and astronomically.

When a class of item qualifies as yellow in the virtual buffer—meaning more than one-third of the buffer is not only missing onsite, but not in the pipeline either—then it means that either very high sales occurred on the preceding day or there are serious delays in resupply (usually in transportation). Several classes of items in the yellow zone of the virtual buffer should generate more pressure to dispatch replenishment immediately. Even one item in the red zone of the virtual buffer should confirm this decision.

Note that if an item shows up in the red zone for on-hand stock only, avoiding a stock-out hangs on arrival of in-transit shipments relatively soon. Normally there's not much one can usually do to accelerate a shipment that's already on the way, but we will review this assumption again later in this chapter.

A red-zone status for on-hand stock could be an important input in a probable decision to increase the buffer, but not really in a decision to transport items to replenish on-hand stocks to the top of the green zone. For instance, if our previous Item X, with a target inventory level of 100, now has only 6 units on hand, this certainly qualifies as an emergency. Stocks could be completely depleted at any time. However, if 75 units are already in transit, nothing is really gained by quickly dispatching the rest of the buffer (100 – 6 – 75 = 19 units) immediately. Efforts are better spent ensuring that the in-transit shipment arrives and is unloaded as quickly as possible.

Rules for transportation are just as important when multiple vehicles transport shipments from a warehouse to the next downstream nodes in the chain. All fully loaded vehicles should certainly be dispatched, but they should be absolutely certain to contain all red- and yellow-zone orders required by the next links, even if it means prolonging the trip or dispatching partially loaded vehicles.

TOC's general rule for distribution is to *report daily and transport frequently.* Determining how frequently is where common sense enters into the decision. Moreover, buffer management priorities imply that expediting is relevant in transportation as well as production.

## Long Transportation Time

Insight 1 of TOC in distribution—shortening the supply response time as much as possible—is all about being very agile in supply. We mentioned the option of delaying shipment when all items in the virtual buffer are in the green zone. But what about very long transportation time? When a shipment takes considerably more than a week (and there are situations in which it could take eight weeks or more), a new question becomes critical: what can we do to ease the pressure of high stock volume the distribution network is forced to carry?

In other words, what changes to the replenishment concept might be required to accommodate long transportation times? The theory is the same: the stock at the end location plus that in the pipeline must cover a liberal estimate of sales during the replenishment time. The liberal estimate also must consider possible delays in the transportation pipeline.

There are two special aspects of long transportation time worth mentioning:

1. The value of frequent shipments in long transportation durations
2. Transporting goods by air when truly necessary

When the shipping time is fairly long, there's even more pressure to ship frequently. Increased frequency shortens the order time. Suppose the shipping time is eight weeks and a ship departs every two weeks. What's the replenishment time? Obviously, it depends on the timing of the reorder, but when the consumption occurs just in time to load its replenishment on a shipment, the replenishment time would be eight weeks. However, if the reorder just misses a shipment's departure, it will be two weeks more, in addition to the eight weeks in transit, before the reorder arrives. Thus, assuming the eight-week transit time is a reasonably stable estimate, the replenishment time will vary between 8 and 10 weeks, with an average of nine weeks. If a shipment departs every week, the average replenishment time would be eight and a half weeks. If there's only one shipment every eight weeks, replenishment time would vary between 8 and 16 weeks, averaging 12.

The frequency of shipments has an even greater impact on the actual level of stock at the destination warehouse. If a shipment occurs every week, the net change in the on-hand stock is the difference between the current week sales and the weekly sales from eight weeks ago (the shipment replenishes what was sold eight weeks ago). Even though this week's incoming shipment is not precisely what was consumed last week, it will approximate what was just consumed as long as demand doesn't peak or dip drastically. The result substantially dampens fluctuations in on-hand stock relative to the somewhat more volatile demand.

The important challenge is still finding a way to reduce the replenishment time. After all, if replenishment time is 12 weeks, the total stock in the system must be sufficient to meet 12 weeks of sales.

When we consider transportation, we need to recognize that there is more than one way to skin a cat. Too often we mentally discard alternative options in favor of a routinely used, relatively inexpensive shipping mode, as if this is the only way. When shipping by sea is the standard, it's easy to lose sight of the fact that air transport can be an option as well. The down side is that flying is much more expensive than surface shipping. But what about value of the mountain of stock one needs to hold at the far end of the pipeline while simultaneously suffering from stock-outs ("Sorry, we're out of that …") and their impact on sales? Can we completely disregard these factors?

It should be noted that the dilemma is not necessarily between shipping everything by sea versus air. It's about deciding to allow some smaller parts of the shipments to go by air so as to be able to significantly cut inventory levels while safeguarding availability. Though regular shipments might be done by sea, target inventory levels can be reduced to some degree by allowing for air transport in case of a really urgent need.

## Criteria for Choosing Air over Surface Transport

Definitive guidance for deciding between air and surface shipping in all cases is probably not realistic. When pipeline times are long, the most pressing need for transporting goods by air occurs when a decision is made to increase the inventory buffer. A signal that this might be required already implies that on-hand stock is dangerously in the red zone. Without air transportation, when the buffer is increased additional stock units would be introduced into the pipeline, but it will require, on average, the full replenishment time for the added stock to arrive at the destination node. In the meantime, stock-outs are highly likely, indicating that the former buffer was not large enough.

Moreover, whenever the buffer is increased there's a period—approximately equivalent to replenishment time—during which monitoring and evaluating for subsequent buffer increases are suspended because the new (current) state hasn't yet stabilized enough to provide valid information. In such situations, only a relatively fast delivery of the net buffer increase would permit a fast subsequent verification of the new buffer size. If we assume that the orders currently in the pipeline move slowly, there will likely be an urgent need (perhaps even an emergency) to move more stock quickly and directly to the destination. When future sales are at risk, the cost of flying should normally be ignored.

What if certain items of on-hand stock are in the red zone? Would this be enough to trigger an air shipment? In this case, a programmed response should be planned very carefully. If the added cost of air shipping is not large compared with the potential loss of throughput, then any penetration into the red zone should trigger air shipment of enough units to bring on-hand stocks above the red level. The rest can make the trip via slower surface shipping. However, if the throughput will be adversely affected by the added cost of flying, a human manager should decide based on specific qualifying rules.

Air shipping is especially important when a new buffer for a new product must be established in a distant warehouse. In this case, the risk of being stuck with a large volume that doesn't sell very well is substantial. On the other hand, if the new product is well received, experiencing sellouts is not a good idea, even if some customers are willing to wait for the product to be available again. Sometimes sellouts can create the impression of a surge in demand, but just as often they represent truly lost sales.

Therefore, for building an initial stock of new products based on a shaky short-term forecast, it may be strategically better to ensure availability by air shipment when a favorable demand develops than stabilize distribution mainly though surface shipment. To summarize, the initial buffer should be based on the maximum forecasted sales within replenishment time assuming air shipment. The build-up of that buffer is accomplished in the course of regular sea shipping. If this maximum forecast turns out to be valid, then air shipments should be adequate to cover for short-term sales, while the bulk shipment required to increase the buffer for the longer, sea-based replenishment travels by sea. So the initial forecast is liberal, but the replenishment time is short. If actual demand turns out not to be as high as the initial forecast, the initial buffer will be adequate when all shipments move by sea.

## Red or Not

Another question specifically related to long transportation time is whether a somewhat different definition of "red zone" is appropriate. Two conditions are necessary for a sensible challenge of the size of the red zone when transportation time is really long:

1. The transportation time must be fairly stable—within a range of not more than plus-or-minus 20 percent of the average time.
2. The frequency of the shipments must be high enough to ensure a relatively stable on-hand inventory at the destination warehouse.

Consider the following example. Let's say that transit time is eight weeks and every week a shipment departs containing backfill for all that was sold the preceding week. This implies that the production cycle is very fast or that shipping is done from stock. The transportation time includes customs processing and varies by no more than ± three days.

Every week, on-hand stock is reduced by the preceding week's sales, which are compensated for by replenishment arrivals equivalent of a week's sales from eight weeks ago. At any given time, replenishment equal to eight weeks of average sales is on the way to destination. We will assume that the target inventory level has been defined as equivalent to 12 weeks of average sales.

Given the stability of the shipping time, and in spite of fluctuating sales but assuming that average sales don't change significantly, a 12-week target inventory level could provide a comfortable safety margin to ensure availability. However, the average on-hand stock will equal one-third of the target level because, on average, two-thirds of the buffer is, at any given time, trapped in the pipeline. This means that at least half of the time, on-hand stock will be in the red zone. Thus, from the red-level point of view alone, the situation appears to be risky all the time, though in actuality the inventory situation is quite stable.

When both conditions apply—stable transportation time and regular, frequent shipments—the red-level threshold should be less than a third of the total buffer, both for setting priorities and as a baseline for monitoring appropriate buffer size. We recommend setting the red level at *50 percent of the average level of the on-hand stock*. The average on-hand stock is equal to the target inventory level minus average sales within the replenishment time.

In the example above, the target level is 12 weeks of average sales. The replenishment time is 8 weeks, so the average on-hand stock is 4 weeks. The recommended size of the red zone could be defined as the equivalent of 2 weeks of average sales. Note that we include a "paranoia" factor of 1.5 in the example. This adds 50 percent on top of the average to accommodate both peak sales and possible transportation delays. The average on-hand stock then includes this paranoia factor.

$$12 \text{ weeks} - 8 \text{ weeks} = 4 \text{ weeks} \times 50\% = 2 \text{ weeks}$$

$$100 \text{ units} - 66.67 \text{ units} = 33.33 \text{ units} \times 50\% = 16.67 \text{ units}$$

The example requires one further clarification. The target level and the red zone are always *expressed in units, not in average sales* because average sales continually change. If the initial buffer value is based on 8 weeks of sales and an additional 50 percent safety factor is appropriate, a target inventory level equivalent to the average number of units sold in 12 weeks is a good starting point. But this number should not be arbitrarily changed (i.e., the buffer increased or decreased) a month later when further data collection and analysis indicates a somewhat different number for weekly sales.

In the preceding example, the red level would be defined as one-sixth of the target level, approximately 17 percent or, in round numbers, slightly less than 20. One additional point: if the red zone is reduced below the default one-third-of-target-level, the possible need to increase the target level requires quick response. And it most certainly dictates the use of airborne shipping for this buffer increase.

## Where Should Most of the Inventory Be Located?

This might seem like a strange question. After all, the whole point of distribution is to bring items to the convenient vicinity of an end customer. Naturally, there are limits to how close to the customer it is practical to put those items. Furthermore, space limitations might prevent storing our preferred volume of items at points of sale (stores). Consequently, most distribution networks require regional warehouses to compensate for space not available at stores. Moreover, stores are usually sited in relatively expensive locations, while regional warehouses are usually in much lower-priced locations.

Figure 8.8 depicts a situation where a manufacturing plant ships finished products to several regional warehouses that supply retailers, or points of sale (POS). The insights and recommendations we have described so far are equally applicable to all nodes in the supply chain—POS, regional warehouse, or central (factory) warehouse. The only factors that determine target inventory levels at each point are (1) the liberal estimate of demand, (2) replenishment time, and (3) the variability range in demand and replenishment time.

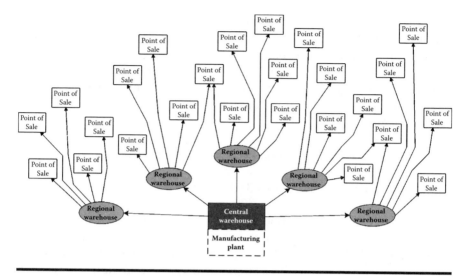

**Figure. 8.8   A typical distribution network.**

## Aggregation and the Law of Large Numbers

But surely location has something to do with it. The point of sale should have relatively more stock than the regional warehouse, shouldn't it? Well, no, not really. Strange as it may seem, the stores should carry a limited amount of stock to meet their need for sales. Understanding the rationale for this should lead us further to conclude that something vital is missing from Figure 8.8. What is typical to a POS is that the fluctuations of the demand are very high while the relative fluctuations of the demand at the regional warehouse are smaller, due to the accumulation of all the POS it serves. The reduction in variation is a natural result arising from the law of large numbers (LLN).

A key concept, the LLN is the fundamental assumption underlying the demonstrated effectiveness of the TOC approach to distribution in dramatically reducing system-wide inventory while simultaneously improving delivery availability. It's important to understand not only *how* TOC-based distribution works, but *why* it works so effectively. The law of large numbers is a theorem in probability that describes the long-term stability of a random variable. Given a sample of independent and identically distributed random variables with a finite expected value, the average of these observations will eventually approach and stay close to the expected value (*Wikipedia: The Free Encyclopedia*, S.V. "Law of large numbers").

## Aggregation and the Effects of Dependence

The law of large numbers assumes that the events being aggregated are truly independent of one another, like the flip of a coin. When such independence pertains, the effect of aggregation on variance can be dramatic. On the other hand, when events (such as a store's daily sales) are dependent to some degree on other factors, such as communication among consumers, media coverage, or targeted or broad-based advertising, the effects of aggregation on variance is diminished. It will never be zero, and the difference can be quite pronounced. In fact, in reality when several stores' daily sales are influenced by common advertising, discounts, or other promotions, the variance range probably more closely describes a continuum, as illustrated in Figure 8.9. The stronger and more direct the relationship among the stores' sales the greater (closer to dependence) the aggregated variance will be.

For example, consider a typical automobile parts store, such as NAPA or A-1. If there is a store in each of two towns about 10 miles apart, it's reasonable to assume that the sales of one will be largely independent of one another. This means that a sale at one will not mean the loss of a sale at the other. A national or regional advertising campaign may increase sales at both of them during the same period, probably at different rates. This means a dependency is created by some factors that impact the sales of both stores. When sales of one store go up, there may be a collateral effect (up or down) on the sales of other stores. Some may change more (or less) than others. A nationwide advertising campaign might be influential for some stores but negligible for others.

On the other hand, if both stores happened to be located in the same town, perhaps no more than three miles apart, it's probable that a dependence relationship *would* exist. That is, a sale at one store would mean the potential loss of the same sale at the other store, owing to the two stores competing for the same customers.

Now, if the same regional warehouse supplies two retail stores in different towns we can reasonably assume that the sales in the two stores are largely independent. The impact of global promotion is much smaller than the independent causes of demand changes in each town, and the net variation in demand at the regional warehouse is definitely smaller than that experienced at each store.

Ideally, we would like inventory management and aggregation of stock to be completely independent because this would produce the most potential for reducing

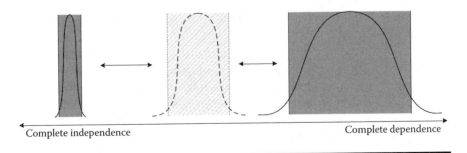

Complete independence                                         Complete dependence

**Figure 8.9   The impact of positive covariance on variance aggregation.**

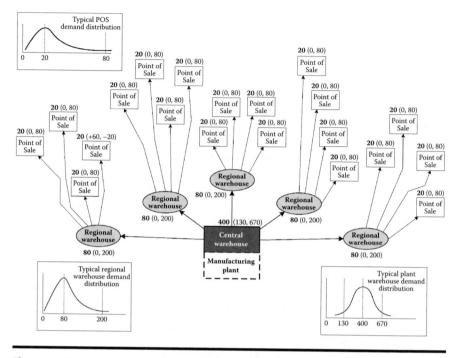

**Figure 8.10 Aggregation of demand and variation in a supply chain.**

total inventory in the supply chain. However, reality lies somewhere in between complete dependence and complete independence. Nevertheless, this still represents a significant potential saving in units of inventory, which in turn translates into much larger financial savings in lost-sale avoidance and sacrifices resulting from scrapping or marking down obsolete stock to move it out.

Thus, it seems that it is advantageous to put more of the material at the region rather than at the POS and compensate by very quick response to the actual sales. When you extend this rationale a step farther, storing most of the inventory at a central warehouse and storing some inventory in regional warehouses that are nearer the retailers, we expect a lower inventory in the total chain. With careful coordination of this system, we can perform with lower inventory while protecting the availability at the retails in an effective way.

Let's demonstrate the effect numerically. Assume that we have 20 retailers that can be evenly distributed into five regions. In each region a regional warehouse can supply four retailers where each retailer has the same size and demand pattern, which is independent of the other retailers. Figure 8.10 provides the potential layout including some numerical values that will be explained below.

At each point of sale, 20 units of product X are sold on average in a day, but demand could be as high as 80 units in a given day. This could suggest a standard deviation of at least 20. While the standard deviation may appear high, it is realistic for store sales and it will be used to demonstrate how the accumulating effect of the warehousing schemes reduces variation.

Each of the 5 regional warehouses supply 4 retailers with a cumulative daily demand of 80 units on average. Suppose the daily demand at the regional warehouse could be up to 200 units in a peak demand day. (While the demand average is linear—simply $4 \times 20 = 80$—the standard deviation could be calculated indirectly by finding the square root of the sum of the four individual variances or the square root of $(4*(20^2))$ = square root of $(1600) = 40$ units [Larson, p. 205]. Note that the standard deviation only doubled for each regional warehouse while the average quadrupled. Thus, the fluctuation level is reduced dramatically due to the integration, if we assume that sales at each retailer are not influenced by sales at other retailers (i.e., the retailers are independent). In any case, the demand at the regional warehouse will show less variance than the retailers' experience because it isn't likely that all four retailers will experience peak sales on the same day. The resulting aggregation impact is based on sound statistical theory.

The same type of reduction in variation can be seen when there is also a central warehouse serving the regional warehouse. The daily average demand that the central warehouse must hold to support 5 regional warehouses is again linear—$5 \times 80 = 400$ units—and its standard deviation is approximately the square root of $(5 \times (40^2))$ = the square root of $(8000) = 89.44$ units, which we will round for clarity to 90 units.

The resulting distribution of demand for the central warehouse averages around 400 units, though on any given day it could be as high as 670 or as low as 130 (average plus or minus 3 standard deviations). The regional warehouses need to hold only enough inventory to cover for the potential high sales during the transportation time from the central warehouse, assuming the central warehouse always holds enough to ensure availability. Table 8.1 summarizes the demand distribution information. The coefficient of variation clearly demonstrates that the variation is reduced from 100 percent of the average when there is no warehousing to 22.5 percent of the average when the central and regional warehousing scheme is used. For more information on the coefficient of variation measure see [Hamburg, p. 197].

## *Lead Time Implications for Specific Scenarios*

Let's assume the retailers are all within one-day drive from their regional warehouse, and that each regional warehouse is within a three days drive from the central warehouse. Further, let's assume the average production time is two weeks. Here are some rough estimates for the inventory in three different scenarios (Table 8.2).

The statistical impact of aggregating daily demand of the various retailers assumes those demands are individually independent of one another. In reality, varying degrees of dependency render the impact of aggregation somewhat smaller. When we expand the aggregation to cover several days, the same statistical aggregation principle applies. The variation in demand at one retail location for 10

**Table 8.1   Daily Demand Distribution by Warehousing Scenario**

| Scenario | For Each | Average Demand | Standard Deviation (σ) | Coefficient of Variation* | Minimum | Maximum |
|---|---|---|---|---|---|---|
| No ware-housing | Retailer | 20 | 20 | 100 | 0 | 80 |
| Regional ware-houses | Regional ware-house | 80 | 40 | 50 | 0 | 200 |
| Central and regional ware-houses | Central ware-house | 400 | 90 | 22.5 | 130 | 670 |

\* *Coefficient of variation reflected as a percentage of the mean. All other values indicated in units of demand.*

consecutive days is much smaller than 10 times the maximum daily demand. How much smaller is not predictable in the real world because of the difficulty in identifying and quantifying all relevant variables. Rough assessments are all we can expect, and these should be adequate.

1. *Scenario A.* The plant supplies directly to the retailers. Replenishment time to every retailer is 18 days, including the production time (14 days) and shipping to the region (3 days) and to the specific retailers within the region (1 day). Average total demand across the retail outlets is 360 units of sales (18 days × 20 units average per day), but demand varies significantly when we focus on a particular retailer at a specific location. Additionally, fluctuation in delivery times could be fairly large. All things considered, we would be well advised to maintain a target level of at least 720 units per retailer. Thus, at any given time there should be 14,400 units in the chain (720 units × 20 outlets).

2. *Scenario B.* The plant supplies to the regional warehouse (no central warehouse). This means that every retailer should have a target level of 80 units to accommodate the extreme demand limit. How much should a region hold? The replenishment time is 17 days, including production time and shipping (14 for production, 3 for transportation to the regional warehouse). The average demand is 1,360 (17 days × 80 units). While the variation is not as wide as it is at a single retailer, the regional warehouses should still add 50 percent of the demand to allow for variability in the replenishment period. So the target level at a region should be 2,040 units (1,360 units × 1.5 safety factor). Because there are 5 regions, the total inventory across all the regional warehouses is 10,200 units (2,040 per warehouse × 5 warehouses). And of

**Table 8.2 Distribution Network Inventory under Various Scenarios**

| | Scenario A | Scenario B | Scenario C |
|---|---|---|---|
| | Plant supplies directly to retailers | Plant supplies to regional warehouses<br>Regional warehouses supply retailers | Plant maintains central warehouse<br>Central warehouse supplies regional warehouses<br>Regional warehouses supply retailers |
| Production time | 14 days | 14 days | 14 days |
| Ship to region | 3 days | 3 days | 3 days |
| Ship to retailer | 1 day | 1 day | 1 day |
| Total days | 18 days | 18 days | 18 days |
| Total retailers | 20 | 20 | 20 |
| Average sales/day (per retail store) | 20 units | 20 units | 20 units |
| Average sales/day (per regional warehouse) | | 80 units | 80 units |
| Average sales/day (central warehouse) | | | 400 units |
| Average sales (during production & transit time) | 360 units (covers 14 production, 4 transit days) | 1,360 units (covers 14 production, 3 transit days) | 5,600 units (covers 14 production days) |
| Sales & transport (variation safety factor) | 2.00 | 1.50 | 1.20 |
| Target inventory level (store) | 720 units | 80 units (for 1 day transit from warehouse) | 80 units |
| Target inventory level (regional warehouse) | | 2,040 units | 360 units |
| Target inventory level (central warehouse) | | | 6,720 units |
| Target inventory level (total distribution chain) | 14,400 units | 11,800 units | 10,120 units |

course, the total inventory in the system, including the retailers, is 11,800 units (10,200 at the regional warehouses + 80 units × 20 retail outlets).

3. *Scenario C.* The plant maintains a central warehouse that supplies the five regions, which, in turn, supply the 20 total retailers. The central warehouse should maintain enough stock to cover *all* the demand in the whole system for 14 days (manufacturing time). The average daily demand is 20 per retailer and there are 20 retailers, so total retail demand averages 400 units per day. Over 14 days, this comes to 5,600 units (400 units per day × 14 days). At the central warehouse, the variation in demand within a 14-day period is relatively low. So let's add 20 percent for fluctuations in the demand and variability in replenishment time: 5,600 units × 1.2 safety factor = 6,720 units in central warehouse inventory.

The regions have to cover all sales for the 3 days of transportation between the central warehouse and the regional warehouse. A particular region sees a demand of 80 units a day (20 units per day × 4 retailers supported). On average, this comes to 240 units for 3 days. As we have seen above, the closer to the retail end we get, the wider the variation in demand, so we add to the regions a 50 percent factor above the average to accommodate fluctuations in demand—and the transportation time, but that should be considerably more reliable. Therefore, the target inventory level at the regions should be 360. The target level at each of the retailers is 80. Thus, the total inventory in the system is: 6,720 (at the central warehouse) + 1,800 (at the regions) + 1,600 (at the retailers) = 10,120 units total throughout the supply chain.

The bulk of the supply chain inventory should be held at the plant warehouse. The rest is distributed throughout the regions and POS based on the replenishment transportation time. The plant warehouse decouples the system from the relatively long—and variable—production time.

In fact, holding inventory at the plant warehouse actually streamlines production. Because the plant supplies a wide range of different end items, without a central warehouse it's exposed to emergency orders originating from everywhere. A stockout of Product X in Region 1 is sufficient to require an emergency order (possibly with premium shipment) to that region alone, even though the rest of the regions have enough of Product X. Of course, at the same time Region 2 may be completely out of another item, of which Region 1 has more than enough of the same item.

Maintaining a central warehouse accommodates these fluctuations, without them ever becoming real emergencies. An emergency order for Product X would occur only when the inventory at the central warehouse dips below the red threshold. The regions may either already have just enough, or whatever a region needs is inevitably already in transit on the way to them.

Most of the advantages of maintaining a central warehouse are relevant also for independent distribution networks not having producers under the same company ownership. In many cases a distribution company's central warehouse, often called a logistical center, stores huge amounts of redundant inventory. Effective management of supplier relationships can reduce the use of excessively large, fully automated, storage space while still securing the benefits of having to store only truly required levels of inventory at regional warehouses and at stores.

The critical lesson of this is:

> Insight 5. Plan the distribution network so that the main bulk of the stock resides at a point central to the distribution nodes, where demand fluctuations are the lowest and the supply chain nodes can be effectively insulated (decoupled) from production time.

## Summary and Review

A distribution function based on TOC principles has three distinct phases:

1. Initial process setup
2. Process operation
3. Smooth modification of the chain layout

At the beginning, the physical configuration of the distribution network might not be ideal, but, for the medium term, it's less difficult to adjust the initial process to the existing situation. Managers of large distribution networks often prefer pilot deployments of TOC procedures before they expand implementation everywhere. Defining such a pilot is a critical step. The larger the number of nodes between a central or regional warehouse to points of sale, the better the results of the pilot will be.

Dispensing with traditional order times, thus reducing replenishment times, requires establishing a data collection mechanism to report daily consumption at all levels. This step alone will reduce occurrences of stockout.

The next step is defining the buffers (the target inventory levels). This can turn out to be a very large number of discrete buffers. Every item at every location should have its own target level. Here are some brief tips for making this potentially tedious work easier, faster, and perhaps not too troublesome.

- Don't be overly precise in setting an initial target inventory level. There is no point in it. The starting figure will undoubtedly be refined before too long.
- Replenishment times should be defined for very large groups of items.
  - These include items that come from the same supplier. Even very different items from the same supplier are likely to have similar actual replenishment time.

- When the replenishment time is mainly transportation from a specific source to a specific destination, then all the items have the same basic replenishment time.
■ To determine the maximum sales within replenishment time:
  - Presuming accurate sales data have been accumulated, a simple program can reveal from historical data actual maximum sales within the replenishment time for each item.
  - A formal forecast can be used. However, caution is in order in deciding on a default rule for "maximum sales." The regular average forecast within replenishment time can be used, or the forecast plus twice the standard deviation or forecasting error.
■ Establishing a safety factor for uncertainty.
  - This is heavily dependent on whether in the preceding steps:
    • The sales figure depicts average sales. If so, the factor should compensate for the "maximum sales."
    • The replenishment time is "reliable," or just an average.
  - The safety factor should be defined for large categories of items. There's no advantage in accepting small differences between items concerning the difference in demand behavior.
  - That said, items that are especially expensive should have a somewhat reduced factor for uncertainty. Items that are very inexpensive could have a larger one.
■ The initial target inventory level for every item/location is the product of average sales during a defined replenishment period and a factor reflecting our confidence in how much higher than average sales could be and in the reliability of the estimated replenishment time.*

It's important to keep in mind that the three-step procedure described earlier should be done only at the initial implementation, when new products are introduced and when a significant change in the sales is expected. The normal procedure for changing inventory target levels should be based on dynamic buffer management and not on recalculating the buffer according to the guidelines described above.

To be able to run the process continuously, a software package capable of following the process must be in place. Existing software applications might be able to support the replenishment procedure. An additional module could be developed for dynamic buffer management.

There is one more prerequisite for ongoing inventory management according to TOC principles. A detailed procedure for kitting, loading, and unloading of the

---

* This is a "whatever-feels-right" number. It might be a factor of 2 or 1.5 or even 3. It represents a starting point that will be adjusted over time, based on observation of what the value of average replenishment time actually turns out to be. However, in our example you might note that our factor increases are very close to the coefficient of variation.

daily transfers throughout the network must be developed. A shipment that supports the agile methods presented here may consist of many items. Each of these may be relatively small in quantity, but they must be collected every day, loaded into a truck, then subsequently unloaded and put on the shelves. A radical change in the number of items per shipment may require significant change in the way items are handled. Effective procedures must be developed to implement and support the new approach.

## References

Deming, W. Edwards, *The New Economics for Industry, Government, Education*. Cambridge, MA: MIT Center for Advanced Engineering Study. 1993.

Hamburg, Morris, *Statistical Analysis in Decision Making*, New York: Harcourt, Brace and World, 1970.

*Wikipedia: The Free Encyclopedia*. Wikimedia Foundation, Inc. Encyclopedia on-line. Available from Internet. Retrieved July 25, 2008.

Larson, Harold J., *Introduction to Probability Theory and Statistical Inference*, 2nd ed., New York: John Wiley & Sons, 1974.

Russell, R. and Taylor, B., *Operations Management: Creating Value Along the Supply Chain*, 6th ed., New York: John Wiley & Sons, 2009.

# Chapter 9

# Managing Raw Materials

## Contents

Managers of supply chains often focus attention on a variety of issues, while paying little attention to their raw materials management. But is there any real difference in the process a manufacturing company uses to purchase its raw materials compared to the process of a distributor purchasing the items to be sold? Certainly, at the conceptual level there would seem to be no real difference. In practice, however, several differences merit careful attention. These include:

1. Distributors usually have a lot of power over their suppliers, which allows them to dictate terms of performance. Some very large manufacturing companies, such as automobile producers, have similar power. However, it is easy to see that small and medium manufacturers have less power to ensure reliable and fast supply.
2. The impact of a raw material shortage may be far more damaging to a manufacturer, because:
   a. It could very well lead to failure in meeting required due dates or unavailability of certain finished goods. Ultimately, serious damage to the reputation

of the company might result. The damage resulting from a missed due date to a client could be much higher than the damage to reputation from unavailability of a specific item at a store. This is especially true when stores keep substitutable items that are acceptable alternatives.

b. Raw material shortages can disrupt capacity in at least two ways. First, shortages can cause starvation of the capacity-constrained resource (CCR), or excessive starvation of a nonconstraint to the point that it emerges as a CCR. Second, when the missing materials eventually show up, a large wave of work may appear on the shop floor, as several delayed batches are now ready for processing at the same time.

c. The missing raw material causes many other materials, already available, to be nonusable for the period of the delay. This may create unanticipated cash pressure at certain times. For instance, when a finished product requires 30 different materials and just one is missing, the other 29 materials represent an investment that can't be materialized even though the opportunity is there.

3. The time horizon for purchasing materials to cover future demand is longerfor manufacturing than for distribution because the relevant horizon in manufacturing includes not only the supply time but also the production time.

4. In some types of production, and for some specially customized materials, ordering of materials occurs only when customer orders are accepted. This means that these materials are not typically held in stock. As a result, the lead time for these materials can substantially lengthen the quoted lead time of a customer order.

5. Distribution companies view effective sourcing as a critical part of running their business. However, in many manufacturing organizations, purchasing is viewed as a necessary evil. The purchasing function is considered to be outside the main line of the manufacturer's business and generally not viewed as a helpful career development assignment. Consequently, the best people aren't usually assigned to deal with purchasing raw materials.

## Availability of Materials

The criticality of managing raw materials in manufacturing organizations varies substantially. The most important determinant of the importance of raw materials management is how badly the production department would suffer from unavailability of materials. In situations where shortages are infrequent, improvement of raw materials management is often concentrated on cost reduction. But this can be a seductive trap. Excessive emphasis on reducing supply costs could easily lead to more frequent shortages of materials, resulting in damage far exceeding the potential cost savings.

In most V- and I-shaped production operations, the availability of materials isn't an issue. In typical V or I plants, the number of raw materials is relatively small. Most of the materials are required for a large quantity of end products, and demand for these products doesn't fluctuate too much. On the other hand, if some products require unique materials having sporadic demand, availability of these items can be a critical issue.

However, in typical A- or T-shaped production situation raw material availability is often a critical problem, because a high number of raw material items are required. A-type plants are dominated by an assembly of many components, each requiring different sorts of raw materials. T-type plants are also dominated by an assembly of various components, but in this case the variety of the components generates a very high number of end products. Both environments require a large number of different raw materials, and any stock-out of even one type of material could bring production to a halt.

## Production Buffers

The use of a production buffer in simplified drum–buffer–rope (S-DBR) assumes complete availability of raw materials. Of course, if a shortage lasts for only a fraction of the production buffer, the brief delay in the timely release of materials may have little effect. If that's the case, buffer management priorities on the production floor will still provide a high probability of delivering on time. However, late shipments from suppliers can easily exceed the capability of the production buffer to offset them. As a result, it's absolutely necessary to buffer the availability of materials so as to ensure meeting market commitments—whether they're firm due dates or commitments to the availability of finished goods.

Later in this chapter we will discuss the scenario in which customized materials are purchased only after receipt of a customer order is received. This rather unique case is much less common, but its impact on the lead time to the customer is significant. In the meantime, we will address materials that should be purchased for stock, meaning there is no firm customer order for them at the time of purchase.

## Purchasing for Stock Based on Forecasts

Manufacturing Resource Planning (MRP) was created specifically to provide relevant information on the required dates for materials. Requirements stemming from firm orders are insufficient to identify all raw material needs. So MRP relies on demand forecasts for finished products to support the timely purchase of materials. But MRP is less than optimal for doing this. Some of the potential drawbacks of the MRP approach are:

- In defining the forecast, demand is almost always expressed as just a single figure. Forecasts often don't include the possible range or variability of the demand.
- The time horizon for the end-item forecasts is expanded to include both the lead time of the supplier and production lead time. This means the forecast must be projected quite far into the future, which compromises the forecast's reliability.
- Because the forecast for end items is, at its very best, an estimation of the average demand, it's especially difficult to deduce the level of safety stock required. The possible high demand isn't included in the forecast information. Why does this make safety stock difficult to determine? A given material might be required for many different end items, making its "maximum consumption" lower than if the material supports just one or two end items. When the safety stock level seems to have no sound basis, the natural tendency is to squeeze it to a bare minimum.

It should be obvious that using end-product demand forecasts to generate raw materials requirements is as poor a way to handle raw materials as it is in managing finished goods inventories. Let's summarize a few insights relevant to effective raw material management:

1. Replenishment time is critical in determining the required raw material stock in the system. Purchasers should strive to ensure fast and frequent replenishment from the suppliers. Purchase orders should be issued frequently and without unnecessary administrative delays. Generally, it pays to devote better managerial effort to purchasing so as to maintain less inventory and better availability.
2. Every raw material item should have a target level, expressed in actual units, that encompasses the inventory on-hand plus replenishment orders already issued, but not yet received.
3. The initial target level should be estimated using *the maximum forecasted consumption within the period of the replenishment time of the supplier, factored by the reliability of the supplier.*
   a. "Maximum" means *more than the average demand.* In other words, it includes an assessment of the required safety. The quotation marks imply that managerial judgment determines how much that safety should be.
   b. A simpler but equivalent approach is *the average consumption within replenishment time factored for uncertainty of both the demand and the supply.* The factor used to determine the target level might be called the "paranoia factor."*

---

* In other words, it's a reflection of the manager's anxiety over running out of stock.

4. Buffer management may be used to draw attention to "emergencies" (near-depletion) in on-hand stock. This would prompt efforts to accelerate the delivery of the orders for that item.

5. Dynamic buffer management (DBM) is used to recommend changes in the target levels. But unlike large distribution systems, in which the DBM could be used to actually change target levels automatically, in purchasing materials for a manufacturing company we suggest limiting DBM to merely signaling the items that the purchasing manager should consider a target level adjustment. Changing the target level must be undertaken carefully, as its financial ramifications can be considerable.

In managing raw materials, the commonly used technique of categorizing the raw material items using an A-B-C scheme may be a useful guide to determine the initial target levels. A items are the most expensive (and possibly most difficult to get) items. These will usually have a lower stock level than B or C items. C items—those that are cheap and common—will typically have a higher stock level to ensure that production is unlikely to be delayed by exhaustion of a C item.

However, buffer management offers a much better way to direct efforts toward what's really required while keeping levels reasonably safe and stable most of the time. The value of categorizing the items can be beneficial for determining target levels, because C items might merit a somewhat higher level because of the "paranoia factor." At the same time, A items could accept a somewhat higher risk of stock-out so as to avoid maintaining too much expensive stock.

# Vendor Relationships

Relationships with suppliers are extremely important to manufacturers. If fast response and reliability of supply can be maintained, flexibility can be realized in facing the uncertain nature of the demand for the end items. When such good relationships are absent, manufacturers pay the price by having to hold relatively large raw material inventories in order to assure effective control of material availability.

One particularly vexing aspect of purchasing is dictation of delivery batch sizes by suppliers. From the manufacturer's point of view, no predetermined batching is best. Purchasing orders should be issued as frequently as possible, coordinated with the actual release of materials to the production floor. The key to maintaining effective material availability without having to hold excessive stock is frequent ordering, elimination of order preparation/submission delays, and dispensing with min–max policies. This is such an important contributing factor that it would be well worth the manufacturer's time and effort to negotiate favorable (and mutually acceptable) ordering and delivery terms with suppliers.

Transportation batching may be considered when the transport cost is significant and batching of several materials within the transport vehicle would save real money, but even then daily reporting and checking should be practiced to facilitate material flow as fast as possible. Batching from the supplier side may be more difficult to handle. Because manufacturers sometimes need to yield to suppliers' policies, it's important to establish that any supplier-imposed batch size should be added to the internally determined target level. For example, let's assume that a manufacturer's raw material target level is 100 and the absolutely minimum batch a supplier will deliver is 75. When the stock in the system (on-hand plus orders already issued) drops below 100, the manufacturer will order 75 units from that supplier. This implies that at any given time the stock of that item—on-hand plus on-the-way—might be anywhere between 100 and 174 units. It should be clearly footnoted that the minimum batch is a blockage to flow that we have to live with until some alternative solution allows us to drastically reduce or eliminate the batch. Another important point: the target level—100 in this example—is the buffer protecting the availability. The additional 75 are only a compromise to deal with high transportation costs.

The basic relationship between a client and supplier is established when the client issues an order for a given quantity to be delivered at a given time. However, the client (the manufacturer in our case) may have only a rough idea of what quantity is required and perhaps an even more vague idea about the time it's required on site. We discussed earlier a client's needs when we addressed make-to-availability (Chapter 6), and we pointed out that MTA also fulfills a legitimate need for most buyers. In considering the question of how to measure supplier performance, this characteristic must be considered as well. Which type of supplier would you prefer, one that's always on time or one that listens to your requests, can accelerate a really urgent order, but might not have the same sense of urgency when a particular order isn't so critical? In the next chapter, we will examine the supply chain as a whole and address the question of assessing the performance of suppliers and downstream nodes.

To be reliable to one's customers, it's absolutely necessary to maintain acceptable availability of component materials. Trying to be optimal by keeping purchasing costs as low as possible could easily result in a ruined reputation. Dynamic buffer management can be invaluable in determining the appropriate level of inventory and successfully managing the ever-shifting balance between consumption and supply.

## Special Orders

Purchasing materials for a specific customer order has very different characteristics than regular purchases for stock. Such a customer order has a defined delivery date attached to it. The production buffer defines the time we would like to have so that from material release until completion we are very much on the safe side. But then there's the delivery lead time (or date) promised by our raw material supplier. Can we rely on it? Suppose the supplier promises to respond to our order in three weeks, but in one specific situation it takes four weeks. This means the supplier has pen-etrated our production buffer by a week. Is our safe delivery to our customer endan-gered? To answer that question, we must know how long the production buffer is.

While our production buffer is adequate to protect against variability in the production process, it is probably not fully protected if raw materials are late. Unavailability of materials could consume too much of the production buffer. Any subsequent visit by "Murphy" might well disrupt commitments to clients. Moreover, while it's possible that the supplier may be late, we realize that we can't rely on the supplier to be early most of the time. After all, if one has a planned completion date, most people see no point in completing early. Thus, most suppliers don't even try to ship earlier than the agreed-upon date. This means that whole supply chain buffer time embedded within the supplier quoted lead time is effectively wasted.

It would seem to be common sense to add some protection time to a supplier's lead time so as not to compromise too much our own production buffer. Because we typically can't expect suppliers to ship early, we might have to identify situations when it's legitimate to pressure a supplier to expedite. Thus, we should add some buffer time to the supplier's lead time, but, of course, not tell the supplier that we actually expect his materials a little later than his quoted delivery date. Instead, we would treat that additional (undisclosed) time as a "red zone," during which we would demand that the supplier expedite our order.

How much should we add to the supplier's time? According to Project Management the TOC Way (Jacob, 1998), it might be reasonable to add 50 per-cent of the supplier's time. This approach is also in line with the buffer management concept described throughout this book that has the red zone constituting 33 per-cent of the total buffer. Adding 50 percent of the supplier's time raises an interest-ing reservation; by having two buffers back-to-back, each of them long enough to provide very good protection, we may have a total buffer that is too conservative. In most cases, we can expect to have all the materials early, and even if they arrive just in time, we still have a full production buffer ahead of us. Wouldn't that be a bit too much? It's obvious that the decision to enlarge the total buffer by 50 percent of the supplier's time is not negligible, and it undermines the pressure by the final customer to reduce our promised delivery time.

In our quest for a simple, effective procedure, without too much protection built in, we've formulated the following principles:

1. The standard lead time for a "not particularly reliable" supplier should be increased by 25 percent. This is our *expected* delivery date (as opposed to the supplier's *promised* date).
2. If the order from the supplier is later than his promised date, it's considered to be "red" and pressure is exerted on the supplier to expedite. Our expectation is to actually receive the order within the additional 25 percent. If it turns out later than that, we still have the production buffer to offset the additional delay.

3. First we calculate our safe-to-promise date based on our manufacturing planned load plus half of the production buffer.

4. From that date, we verify whether our customer's due date – production buffer – supplier's time × 1.25 produces a date that is already past. If not (this could likely happen when the existing planned load is fairly long), then our safe-to-promise date is acceptable. However, if the above calculation produces a date that's in the past, we need to extend our safe-to-promise date until it's precisely production buffer + supplier's lead time × 1.25.

Here's an example:

■ Our planned load is now 6 weeks.
■ Our production buffer equals 4 weeks.
■ Our supplier's lead time is 4 weeks.
■ Our initial safe date (not considering the supply) is: 6 + (4/2) = 8 weeks.
■ Testing the validity of the safe date: 8 – 4 – (4 × 1.25) = –1 week.
■ Because this date is already behind us, the real safe-to-promise date is: 4 × 1.25 + 4 = 9 weeks (the same as adding the expired weeks to the initial safe date).

Note that if we issue the purchase order today, the supply will arrive—we hope—in four weeks. But we're prepared to receive it as late as five weeks, still leaving our four-week production buffer intact. Naturally, if the materials actually arrive in four weeks, as the supplier promised, we have an additional week of slack in the production buffer. If the supplier is late by an additional (sixth) week, the order will be released to our production floor three weeks before the delivery date—still "in the green" (a penetration of one week into a four-week buffer is still in the green zone).

To reiterate, the situation in which material is purchased specifically for a customer order is an exception rather than the rule. In almost all cases, raw materials will have to be purchased to stock because most customers won't tolerate excessively long promised delivery dates. Special orders might be an exception to this rule.

The ideal Theory of Constraints (TOC) supplier is obviously one who is willing to assume responsibility for our raw material stock himself and provide us with assured availability. However, the supplier also needs to win from such relationship. This kind of win–win agreement is at the heart of the TOC vision for supply chains covered in the next chapter.

# Reference

Jacob, D., Introduction, *Project Management the TOC Way—A Workshop*, New Haven, CT: The Avraham Y. Goldratt Institute, 1998.

*Chapter 10*

# Business Practice and Supply Chain Performance

## Contents

The concept of supply chains and supply chain management—a decidedly whole-system perspective—has achieved wide acceptance for modeling a variety of businesses. What probably makes the term "supply chain" stick in many managers' minds is the recognition that any link in the chain prospers over the long term only when a customer buys an end product. Why this is such a recent phenomenon we might question, but its global acceptance is undeniable. How can we explain this broad acceptance?

First, the chain analogy is easy to understand. It's also certainly valid over the long term. Suppose that you are a producer of the microchips used in notebook computers, and you have sold a large order of these chips to a notebook original equipment manufacturer (OEM), such as Apple, Hewlett-Packard, or Dell. If the OEM installs your chips into the notebook computers, customers buy them, and, if they like the end product, demand increases and everyone benefits.

Alternatively, you might sell a large order of chips to an OEM who runs into financial difficulties and doesn't complete the production of the laptops. Obviously, there are some situations in which someone in the supply chain sells to the next link and receives payment, but the product never reaches the final user. A benefit to the supplier, for sure, but in the long term it's a one-shot deal. Actually, this is frequently a risk in high-tech industries. Because technology and new high-tech models evolve very quickly, supply chains may become filled with so much merchandise that a substantial amount of goods won't sell at all, while another large percentage may be sold, but at significantly reduced prices.

Why should a microchip producer care whether an end user has actually bought a particular notebook computer when their chips are only a tiny part of the end product? Managers at microchip fabricators should be very concerned. It's crucial to the efficacy of the concept of supply chains. Whatever is sold to the next link in the chain must be truly sold because future sales ultimately depend on the success of the downstream links to sell continually—to "pull" more inventory from the supply chain. Consequently, the logical linkage between selling to successive links and the final sale to the end consumer is very strong indeed. It's critical for any producer to support the sales of the final consumer product.

## Supply Chain Members Sink or Swim Together

The performance of distributors and retailers is linked to the performance of producers. Even if a distributor astutely uses Theory of Constraints (TOC) buffer techniques to ensure reliable availability, an unreliable, slow producer typically negates a major benefit of TOC by forcing downstream links to maintain unnecessarily high stock buffers. When an entire supply chain holds too much stock, some of it is sure to be in slower-selling items—those that, if only you had known ahead of time, you would have ordered very few (or none) of them.

Nonproductive stock like this limits investment in stocks of faster-selling items that could be more profitable.

The effect is compounded at each link in the chain. In a chain that holds too much stock, all nodes eventually suffer. Even the producers of slow-selling items are impacted because these items are "clogging" the warehouses. While they may have sold adequately upstream (prior to the distributor), the producer subsequently won't be able to sell any similar stock it's currently turning out. The unfortunate producer might have been able to shift to faster-moving products had they only known their slow-moving products were jamming the supply chain. But it typically takes a long time after production for the real demand to be revealed through sales. This delay often frustrates the agility required to come up with new products that will sell immediately in the marketplace. Moreover, even if such information resides somewhere in the supply chain, without a concerted effort by the individual links to communicate among one another, that information never gets to where it could do the most good.

Managers of supply chains commonly believe that lack of information is the problem. Thus, their remedy inevitably is to generate more information, to reach out for new communication technology, and to share information throughout the chain. However, is this really the problem? Is it merely a technical problem that, once solved, will promote information sharing and speed response throughout the chain? Sharing information among supply chain members is crucial, but it's not a panacea. Consider the following example.

A distributor we will call "HT-Sales" realizes that it's carrying too much inventory in a particular model of notebook computer made by LapOne. The computer has been selling very slowly. Now, LapOne is developing a new model, which will be lighter in weight. It's predicted to have excellent demand. So here's the practical dilemma: should HT's sales department tell LapOne that units of their current model are moving like molasses up a tree on a cold day in January? What good would it do? LapOne might accuse HT's sales department of not promoting their products effectively. This might prompt LapOne to strengthen its business relationship with a competitive chain, especially for their new model.

If HT's salespeople don't tell LapOne about it, LapOne may decide to wait for HT to provide a forecast of its new model. Because HT's sales force is closer to the market, they're more likely to be able to reliably forecast demand for the new model. This could be very useful information because it might indicate how much capacity LapOne should dedicate to the new model and how much initial stock to build. But what are HT sales' own interests? To assure that LapOne can meet a sudden large demand for its new model, it might be a good idea to "pad" the forecast a little. As long as it doesn't require committing to ordering a very large quantity, increasing the forecast might well line up with HT sales' interests. But it's absolutely *not* consistent with LapOne's interests.

Whenever two companies collaborate, their respective individual interests must be carefully considered. A supply chain creates a unique association between various organizations. Their business transactions must truly support a win–win partnership. Only such relationships will lead to the optimum performance of each individual partner in the chain in a way that benefits the whole chain.* The relationship should foster agility in adapting to the changing tastes of the market while simultaneously avoiding too much risk.

## Inventory Levels Are Key to Vitality

Eventually two critical capabilities are required for a supply chain to prosper:

1. Speed in launching new products.
2. Offering wide variety with assured availability at every point of sale.

Some firms may find that the quick launch of new products makes it difficult to offer variety and availability at all retail outlets. Usually the reason for this is the perception that maintaining availability of a large variety of items requires high levels of inventory of all items. Large fluctuations in demand, especially when many items compete for consumers' attention, exacerbate this problem. When new products are introduced, most often to replace older items, the investment in inventory multiplies. Rather than balance or trade off the two critical capabilities, companies persistently try to expand or improve both of them. In order to achieve both capabilities, above, the chain must be capable of:

1. Maintaining a low level of inventory *throughout the whole supply chain*.
2. Fast replenishment of anything that is sold *anywhere* in the chain.
3. Maintaining availability of materials and products at *each link* in the chain.

The logic we have presented throughout this book supports the assertion that low inventory levels are required to achieve both critical capabilities. In order to provide a fast launch of new products, older products must not block this launch. When the remaining inventory of older products is significant, some links in the supply chain—those that own the stock of the older products—will tend to resist introducing new products. It isn't possible to overrule these objections every time new products are introduced, nor is doing so a "win" for all players in the chain.

Low inventory levels are necessary to maintain the availability of a variety of items as well. Otherwise, capacity is wasted by producing and storing what isn't immediately necessary. Remember that cash and space are common constraints with distributors and retail stores, respectively. Production companies don't normally have these limitations, but they're constrained by their capacity-constrained

---

* In the immortal words of Benjamin Franklin, "We must all hang together, or assuredly we shall all hang separately."

resources (CCR). This is especially so when making to stock (MTS) based on a forecast. Additionally, an obsession with efficiency often drives manufacturers to try to reduce allocated unit product cost by producing large quantities. Manufacturers who often find themselves running out of capacity actually waste it by producing items that won't sell.

In order to maintain low inventory without disrupting availability, all upstream links must be responsive to any sale from the final node in the chain—the point-of-sale. How can the lead time through the entire supply chain be compressed? Every link is an organization with its own interests. The physical location of some links might be far from adjacent links, meaning transportation time is not negligible. If some links unilaterally take unwarranted risks to overcome the effects of long distribution intervals, available production capacity may be exhausted. This can be especially problematic when CCR capacity is fully consumed by MTS orders that aren't really required—they're just being completed to make arbitrary efficiency measurements look good.

Achieving the quickest response from the chain as a whole requires execution of a scheme that allows each link in the chain to maintain full availability of its products to the next link. Such a scheme is essentially similar to a *kanban* system. But in this case its purpose is to decouple each link from production, and from the transportation time from its preceding link.

The TOC approach is to combine making to availability (MTA) with the methodology for distributors and retail outlets described in Chapter 8, and to do so throughout the entire supply chain—including premanufactured raw material production and delivery, if possible. The manufacturing company takes responsibility for ensuring the availability of its products at the client's (distributor's) site. Distributors, in turn, offer the same assurances to retailers, thus encompassing the whole supply chain.

Such a supply chain achieves both agility and superior availability. Producers assure relatively low stocks at the next link while ensuring good availability. Distributors quickly replenish their retailers. And everyone achieves both agility (constant updating of offerings to the market) and excellent availability. Strictly controlling the entire supply chain with effective buffer management reduces the total investment in stocks throughout the supply chain while guaranteeing the highest levels of availability. Hence, inventory turns—the prime distribution measurement—are kept high. This is highly advantageous to the supply chain, and availability is enhanced as well.

The concept described above (and illustrated in Figure 10.1) is not easy to realize. However, the difficulties lie not in the technology or the procedures, but in the human psychological component. The TOC supply chain concept is a fairly radical departure from accepted ways of managing production and distribution, which typically treat each element of the supply chain as isolated links. Successful creation of a robust TOC-based supply chain requires close, effective coordination among the links. Deming has observed that the most acute problems in organizational

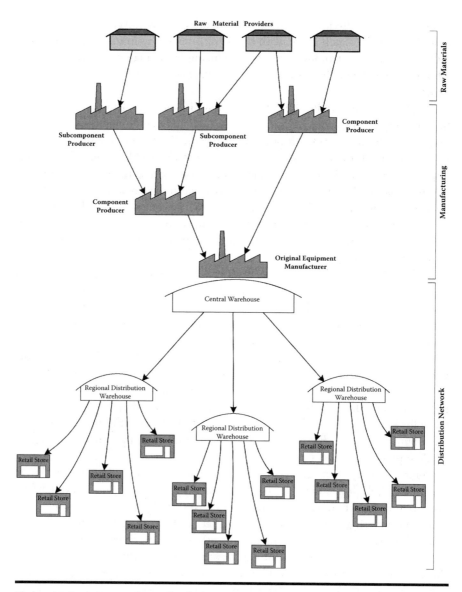

**Figure 10.1 Integrated supply chain.**

chains reside in the linkages, or interfaces among links, not within the links them-selves. Nowhere is this more obvious than in supply chains, especially the ones composed of different companies, each with its own goal and objectives.

Thus, the idea of win–win relationships becomes absolutely crucial. Each link in the chain must see itself as part of a team with the others. It's clear how the supply chain must function as a whole to maximize business. But to elicit the cooperation required for success, each link must perceive the benefit to itself in doing so—in other words, they must see "what's in it for me." In order to ensure that each link in the chain wins from employing the TOC methodology, we must examine the most common business practices to see how well they support the TOC scheme. Do they really promote win–win relationships throughout the supply chain?

## Common Business Practice

In normal business practice, clients specify what quantities of each item they need. Together the client and the supplier agree on a price and delivery date—a simple, straightforward mechanism based on mutual understanding. What could possibly go wrong? Under this scheme, the client assumes the risk of not being able to sell what he bought—whether he sells the supplier's product as is, or uses them as materials in his own production. Obviously clients want just enough, but not too much. How should clients decide how much to order?

Now, the client is closer to the market and might have better intuition, or perhaps even substantially more knowledge of what items might be in high demand in the short and medium terms. Thus, it seems reasonable for the client—the downstream link in the supply chain—to determine the quantity to be produced. But as we've seen in Chapter 8, the quantity to be produced is also related to the resupply time. Who knows better how fast and how frequent this product can be replenished? Considering this question, the supplier has a substantial advantage.

How much effort should a given link in the supply chain make to be responsive to the real needs of the subsequent link? If that subsequent link decides the quantity to be ordered, suppliers naturally want to take for themselves as much time to deliver as their business relationships will allow. If the client is aware of a relatively long replenishment time, then the order will tend to be relatively large, considering longer scope and, therefore, less accurate forecast. In situations like this, suppliers can be satisfied only for the very short term because they feel that they have little or no control over what's really good for the supply chain as a whole. The need to shorten the response time of the supply chain, and the ensuing clash of interests that often results, is related to the conventional approach to business. The absence of a win–win solution has led some of the very large companies in the world to move toward substantially different business practices—for example, vendor managed inventory (VMI).

VMI implements a basic scheme that we believe is good for the supply chain, namely that the supplier takes full responsibility for the availability of his own products at the site of the client. This method puts the risk on the shoulders of the supplier. As a result, the supplier now desires to be *really* responsive to the client's consumption of his products. But VMI, particularly its win–win characteristic, raises many questions about business practices. Large companies, such as auto manufacturers and huge distribution networks, have imposed VMI involuntarily on suppliers. Naturally, their suppliers don't see VMI as being good for their own businesses. From their viewpoint, the shift to VMI was definitely a win–lose kind of relationship.

Of course, within their limited power suppliers inevitably try to play their own game to realize an advantage whenever they can. Since the client is closer to the market, and in this kind of "power game" has the final word, changes in the forecast are noted by the client, who relays them to the suppliers. This is actually one of the soft points in this kind of win–lose relationship because, when the client renders his forecast, he also commits to the forecast to a certain extent. Consequently, in most VMI agreements the clients take the responsibility for the forecast out to a certain time horizon, and if a decision is received to stop selling a certain product, the supplier is compensated for the amount of the forecast that is obviated for the defined scope of time.

Generally, establishing win–lose relationships doesn't facilitate a real breakthrough in performance. For example, the automobile industry has long ignored a market that wants fewer gas-guzzlers and more fuel-economical cars, to its own detriment and that of upstream supply chain members. Although suppliers met the VMI requirements in the short run, the erroneous decision made by the powerful automobile manufacturers has compromised their own "win" component and exacerbated the "lose" aspect of their supplier relationships. It would be far more preferable to adopt a concept that still employs VMI, but which is based on true win–win relationship before it deteriorates to lose–lose. The methodologies described in this book are potentially beneficial for such a strategy. However, what principle underlying business practices will support such a win–win scheme?

It's not difficult to recognize lose–lose situations. Consider Super-Mart, a major retailer enforcing a strict requirement of rejecting late deliveries. If a shipment arrives half a day late and Super-Mart refuses to accept it, everyone loses. Super-Mart is short of inventory, the trucking company is caught in the middle, and the manufacturer loses a sale as well. Selling the products at a discount off the parking lot isn't really an option. It competes directly with Super-Mart and adds selling costs for the manufacturer and/or trucking company. What other options does the supplier have? To be on time, all the time? What happens if, despite all efforts, a particular order is late? Isn't shipping the order anyway still a valid option? The power of the retailer to impose such a rigid requirement on upstream suppliers—and strictly enforcing it—hurts all parties. A supply chain can't survive this type of relationship.

Once the inherent dependency among supply chain members (links) becomes apparent, it becomes clear that partnership protocols are essential to effectively configure business relationships. This isn't easy to do, especially because the members of one supply chain might also be involved in other supply chains that may compete with each other. But in order for a supply chain to realize the capabilities required for breakthrough performance, *all* links must be motivated to support the capabilities. And for that to occur, everybody must be rewarded based on the real gain for the chain as a whole. True partnership implies that each link in the chain makes money when the last links sell. Prior to such a sale, not all links can make money. The primary objective of this approach is to apportion the throughput from each sale among all the participants in chain only after the final sale takes place.

Throughput (T) in a for-profit organization is defined as: *revenue minus the truly variable costs per single sale.* The term "truly variable costs" means that whenever a sale takes place, these costs occur as well. Normally, we consider the raw materials included in the products sold as truly variable costs. This is certainly true when replenishment procedures are followed, because any sale causes replenishment in producing the items, which causes replenishment of raw materials required for the quantity of the sales.

New ideas have many ramifications. Evaluation of any idea should go through three phases. An analysis of the new idea should:

1. Show that the idea has real benefits.
2. Raise reservations (negative branches) showing any adverse consequences that could result from implementing the idea.
3. Seek an updated solution that would keep all (or most of) the benefits and get rid of most of the negatives.

Once an idea has cleared these three phases, obstacles to implementing it must be identified. For example, would a new computer system be required to manage transactions throughout the entire supply chain?

## *Benefits of Apportioning Throughput*

Distributing the total throughput achieved from every sale among all participants that have contributed to the sale suggests a high likelihood of promoting win–win outcomes for all supply chain members. Let's demonstrate this idea. Suppose LapOne produces a computer called Lap1. The truly variable costs of producing a Lap1 are $450. Super-Mart sold one Lap1 for $999. The full retail price was $1,150, but Super-Mart opted to offer a $151 discount. So the throughput realized by the LapOne—Super-Mart partnership is $449. If the two supply chain partners agree that LapOne should receive 40 percent and Super-Mart 60 percent, then immediately following the sale LapOne earns its $450 variable cost plus $179.60 (40 percent of the net throughput), or $629.60 total. The benefits of distributing throughput this way include:

■ Establishing the concept that any link in the chain makes money only when a product is sold to the end customer thus tying supply chain members together.
■ Flexibility for the last link in determining the selling price, taking into account the real throughput that the sale would generate. Traditionally, stores

have been held captive by the prices they have already paid to the supplier, which contains also the supplier's predetermined throughput.

■ Much better information for all participants concerning how their products, or product components, are selling.

■ Consideration of the throughput generated by the chain in decisions any supply chain link contemplates. This is likely to promote better alignment between local decisions and the success of the supply chain as a whole, assuring that the win–win approach is actually realized.

## Obstacles and the "Law of Unintended Consequences"

The favorable ramifications of distributing throughput from each sale are considerable. However, there are also some potential obstacles and negative outcomes. Let's talk about the obstacles first.

If throughput is divided among the partners after the sale, then until that event the production, transportation, and storage of components and products represent an investment that each link is required to make. Now the farther upstream a link is, the greater the patience required in waiting for a return on this investment. Moreover, unless otherwise negotiated, manufacturers would have to invest in both the materials *and* the production capacity, and wait for some time until they can expect payment for both the truly variable costs invested and their part of the total throughput. This perception could discourage supply chain members from even trying the TOC-based solution in the first place.

Verbalizing our concerns helps to think about an appropriate solution. First, let's examine the issue of the time it takes until payment is made. Historically, manufacturers have not been paid immediately upon delivery of the goods. Business terms can delay payment for 30 to 60 days, or even longer. If a supply chain is really based on fast replenishment, and if the distribution of throughput among those who participate in its generation is very fast—as fast as the TOC supply chain solution promises—then payment times will be comparable, or perhaps even faster than payment terms commonly employed today.

Let's assume that the overall concept contains the seeds of the solution to this obstacle, as just described. What if the time duration of the investment turns out to be a real issue? Moreover, another negative branch is linked to it: How does such an integrated supply chain handle products that don't sell, or are sold at a loss? Who should carry the loss? If we expect to have a true partnership, the partners ought to share the losses as well as the profits. What operating rules might be established to answer these concerns?

Our basic assumption is that a supply chain that's truly fast (i.e., reacting to the rapidly shifting preferences of the market and providing excellent availability of whatever it offers) can be very successful. It can be even considerably more successful than if it's managed conventionally. If this is true, the additional profits earned by the links of the supply chain are more than enough to add another partner—a financial partner capable of providing financing for the whole chain. This can lessen the impact of excessive time between resource commitment and compensation for supply chain members. It could help prevent losses as well. As revenues come in, the selling partner collects his share of throughput and pays all the other partners. Each partner must still invest in its own capacity, but that should be considered a normal part of doing business.

What we've presented so far is a broad vision for managing the supply chain. The operational part is largely a combination of the previous chapters. The global objectives, the business practices and the way to institute the win–win approach have been briefly described here, but they don't constitute a fully developed solution yet. "The devil is in the details," and the general concept will require tailoring for specific circumstances. But the direction we've presented should be quite clear.

## Conflicts of Interest

Our suggested approach offers a new, significant opportunity to align the interests of the various partners in the supply chain, but it can't fully guarantee that this alignment will happen. To ensure a win–win culture is institutionalized, and to constantly search for improved answers, we must do two additional things:

1. Overlay reinforcing performance measures on the various links along the chain.
2. Promote and develop a culture that encourages dealing with conflicts of interests.

Let's address the second one first. Selected performance measurements will be discussed below. What situations give rise to such conflicts, and what assets do we have available to support conflict resolution or management? Consider, for example, these potential conflicts:

■ A new and innovative product is introduced. The main producer feels the old allocation of throughput doesn't cover his extensive research and development expenses. (The supply chain won't cover these expenses because it has no exclusivity over the useful life of the product.)
■ Two or more producers compete with very similar products.
■ Divergence between the priorities of the entire supply chain and an individual "link." For example, let's say a producer has a bottleneck at a certain link. The producer's priority, determined through its internal calculation of throughput per unit of its constrained resource's time (T/CU) might be very different than the priorities of the chain as a whole.

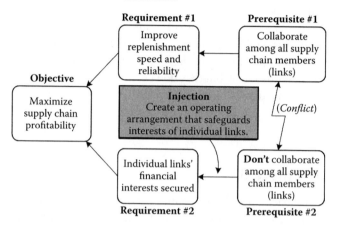

**Assumptions**
1. Collaboration is absolutely essential for speed and reliability improvements.
2. One link can compromise the whole supply chain.
3. All supply chain members can be trusted to act for the greater good.
4. Rapid communication is essential for speed and reliability.
5. Rapid communication doesn't happen without effective collaboration.

**Requirement #1** — Improve replenishment speed and reliability

**Prerequisite #1** — Collaborate among all supply chain members (links)

**Objective** — Maximize supply chain profitability

**Injection** — Create an operating arrangement that safeguards interests of individual links.

*(Conflict)*

**Individual links' financial interests secured** — **Requirement #2**

**Don't** collaborate among all supply chain members (links) — **Prerequisite #2**

**Assumptions**
6. Existing ways of operating are tolerable.
★ 7. Other links can't be trusted to operate for the greater goods.
★ 8. The potential benefits of collaborating don't outweigh the adverse effects of losing control of financial interests.
★ 9. The risk of losing control of our future increases with collaboration.

**Figure 10.2  The supply chain collaboration conflict.**

■ One link in the chain has serious financial difficulties, which might cause a major problem for the larger chain.

## Evaporating Clouds and Negative Branches

Our point here is not to offer detailed solutions to these examples of typical supply chain conflicts that might arise, but rather to offer a process for solving these problems in a way that constitutes a win for the entire supply chain and each member of it. TOC provides two tools that are specifically built to deal with these situations. The first is the conflict resolution diagram, also known as an "evaporating cloud" (EC). The EC clearly exposes the essence of the conflict among supply chain links and helps bring to the surface the basic assumptions that actually produce the conflict. Challenging these assumptions is a sound first step in creating breakthrough win–win solutions. Figure 10.2 illustrates an example of a typical supply chain

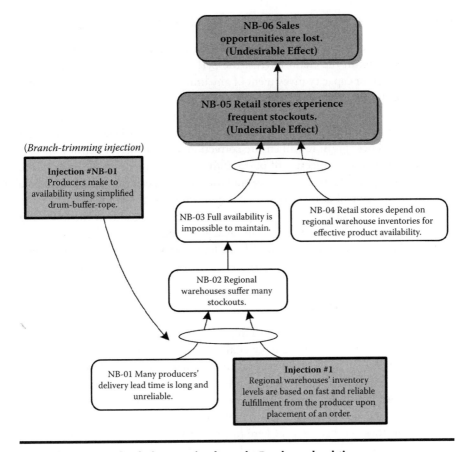

**Figure 10.3  Supply chain negative branch: Producer lead time.**

conflict structured into an evaporating cloud. For a more comprehensive discussion of evaporating clouds, see *The Logical Thinking Process: A Systems Approach to Complex Problem Solving* (Dettmer).

The second tool—the negative branch—is crucial in verifying that a new idea for a solution doesn't create new, devastating problems (possibly worse than the one it's intended to solve). The negative branch helps answer the question: "Besides the outcome we hope for and expect, what else might our solution do that we don't like (or can't stand)"? The negative branch procedure also suggests ways to head off the undesirable consequences before they even occur. Figure 10.3 illustrates a negative branch related to the effect of producers' lead times on availability. (See Dettmer, *The Logical Thinking Process*, for more on negative branches.)

# Performance Measurements for the Supply Chain

The concept for a "warp speed" supply chain described above, and in Chapter 8, requires each link in the chain to trust the performance of the other links. After all, the financial or capacity investment of any link will prove fruitful if—and only if—all other links also do what they're required to do. Even in a win–win culture, inferior performance of any link, for whatever reason, must be immediately obvious to all the other links. Once the other links discover the deficiency, contractual obligation and peer pressure will ensure correction of the deficient performance.

Consequently, the objective of performance measurement for all links in the chain is the rapid detection of any degradation in the performance of the whole supply chain. Remember Ben Franklin's admonition. Upstream links must ensure that the availability of materials, components, or finished goods at the following link is at or above the standards set for acceptable performance. Measuring the performance of the downstream links has a lot to do with moving the stocks fast enough through those links. We'll discuss upstream and downstream separately.

## *Upstream Performance*

The concept we offer for integrated supply chains ties the local decisions of every link to the performance of the supply chain as whole. Performance of individual suppliers is measured by what actually occurs, not what they agreed to do. If a supplier takes the necessary precautions to prevent stock-outs, his measurement (and reward) should reflect it. The agreement between client and supplier is important for establishing expectations, but is less useful in evaluating performance.

When a stock-out occurs anywhere in the chain, it's clear that something has failed, although *real* damage, such as losing sales, may still be avoided if the stock-out is far from the selling points and the planned precautions prevent the "bubble" from ever reaching the end customer. Nevertheless, any stock-out anywhere in the chain poses a risk of lost sales. Consequently, we should concentrate on measuring the *availability* of items, where the responsibility lies on the supplying link. If damage—lost throughput opportunity—actually occurs, it's proportional to the throughput of the end item. The duration of the stock-out or shortfall is also critical because each additional day that the failure-to-provide exists there's a greater risk of a lost sale.

Note that we can't assess real damage here because it's not known. Moreover, the damage is not an actual out-of-pocket financial cost because the missing product will likely be sold eventually, once the "hole" is filled. The potential lost throughput is an *opportunity cost*.

Even when we consider a stock-out at a store, we don't really know how many sales, if any at all, might have been lost because stores don't usually track instances when customers ask for something that was out of stock. Moreover, customers often will not ask anyone; they will simply look, and if they don't find what they want, they simply leave the store. Therefore, it's unlikely that an accurate figure for potential lost sales can ever be known. But the hidden damage to a business's reputation resulting from a stock-out could be much more significant. How can one measure the impact of this customer attitude: "I don't shop there anymore because, as often as not, they are out of what I'm looking for."

We shouldn't waste any time or energy searching for a precise measurement of actual damage resulting from a stock-out. This really isn't even the objective. All we want is to know whether the performance of a link in the supply chain has begun to deteriorate. That should be more than adequate to convey the potential of possible damage, even if we can't measure its actual impact on the supply chain. Really, we must be able to define something much more functional: *guidelines for making appropriate choices when we are faced with inevitable shortfalls.*

## Loss of Throughput

For every shortage, we will certainly factor in its duration the throughput of the threatened end item, and the average daily consumption of the item. We do this so as to obtain a rough assessment of the potential damage. The total potential damage is added to a periodical measure—a monthly measure of lost throughput—of the failure to maintain availability for each supplying link in the chain.

The potential damage is measured in *throughput value days*.\* It's determined by multiplying the throughput per item by the average consumption of items per day, then multiplying again by the number of days the item is out of stock or "in the red zone."

The throughput per item metric requires some clarification because determining throughput per item is complicated by a couple of situations. Throughput is generated by selling finished products to end users. However, many different finished products might use the same raw material, which might experience a shortage or stock-out. The conceptual challenge lies in allocating the effect of a raw material shortage to the end item of interest. Availability of raw materials, which might cover a broad range of end products, is one basic criterion in the approach proposed here. Additionally, the relationship between raw materials and end products isn't necessarily one-to-one. Sometimes the ratio may vary as a result of scrap or other forms of waste during manufacturing. Existing standards that determine how many units of material X are required for a unit of end product Y sometimes overlook this issue.

The important point is that we don't need to spend an inordinate amount of time coming up with detailed formulae to deal with these complications. It's sufficient to assign a "typical throughput" value to every raw material item considering the end products it is part of. Also, we should avoid changing the data (typical throughput) each time end product price is updated; just determine a value and stick to it.

---

\* "Value," in this case, is the actual financial currency used in the situation: dollars, pounds sterling, shekels, yuan, yen, pesos, etc.

An approximate assessment of lost throughput allows us to recognize quickly when a supplier's performance begins to falter. It's a kind of measure that rises sharply when a material item is short for more than one day, emphasizing the fact that the damage of a shortage or stock-out is far above the cost of the item itself. The metric itself has all the ingredients of throughput value days that we apply in make-to-order firms—it focuses on the final throughput and the duration of the delay. Its sensitivity allows for rapid identification of a problem.

It's not critical to identify all shortages instantaneously. But we don't need to wait for a monthly (or weekly) report when we have buffer management, which virtually "screams" whenever a particular product goes into the black. So why do we need the throughput value days to signal supplier performance problems? In reality, we will never have perfect availability. Here and there some shortages or stock-outs will occur, but maintaining huge inventories to avoid such low incidences of shortages isn't really practical. Rather, it would be preferable to find a supplier able to provide very good, maybe even excellent, availability—one who assumes the responsibility to cover any shortfall very quickly to limit the loss of throughput. In a supply chain that delivers many thousands of stock-keeping units (SKUs), a certain number of stock-outs is inevitable. The challenge is to keep them as few as practical, and we want to know when that number starts to grow. To identify deterioration in supplier performance, we need information for every supplier's items and the record of shortages and durations. The throughput value days metric identifies actual shortages (or stock-outs) and their time span. Though they do exist, they occur only sporadically. However, this sporadic occurrence causes two control problems:

1. The throughput-lost assessment graph may have too few data points to convey a clear view of whether performance is actually deteriorating or we are merely experiencing a random peak. Some months might not experience any shortages, while other months might have two or three "red" or "black" items causing a sharp peak in the graph.
2. Once deterioration is clearly identified, significant damage may have already occurred if a considerable number of stock-outs have already occurred. Our objective is to identify such deterioration very quickly. It's not clear that problem identification based on a throughput-lost metric is responsive enough to correct a potential deterioration before damage is really done.

## Throughput Value Days of Red Penetration

Because we want to be sure that we detect deterioration in supplier performance before the damage to throughput occurs, let's test another approach—measuring the throughput value of what is missing in the red zone. The concept is similar to dynamic buffer management (monitoring penetration of the red zone). But in this case, we want to combine red-zone penetration with the throughput value of the end product in order to give proper weight to the damage that could have been expected had the red-zone penetration been a real stock-out.

Throughput value days (TVD) of red-zone penetration is calculated daily and added to a periodical counter, such as a monthly counter per supplier. Here is how the calculation is determined:

- Every day, for each supplier, the supplier's items that penetrate the red zone at the next link are tallied.
- For each such item, the number of units missing (in the red zone) is multiplied by the assigned typical throughput per item unit and added to the supplier's tally.

**Note:** If the item is actually short then the whole quantity of the red zone is considered plus average daily consumption for that item.

The benefit of this metric lies in its sensitivity to two variables: (1) the duration of a shortage (the penetration of the red zone), and (2) the value of what is short (red). Any real divergence from the reference performance would be revealed almost immediately. In any month, several items may be expected to be in the red zone. After several months a statistical mean can be established, along with control limits. Thereafter, any deviation from the upper control limit should signal a warning that the supplier's performance is deteriorating. Dialog with the supplier must then be initiated.

A disadvantage of this measurement is that it states a value in dollar days—units that do not feel "real," because it measures damage that hasn't really occurred. This happens because most people are not accustomed to using dollar-days. The throughput-loss assessment presented above is somewhat more realistic, even though it is not a precise metric. We recommend using both. The "TVD of red penetration" metric is the most useful for avoiding stock-outs, while the throughput-loss calculation can identify the costs of situations where damage has already been done.

## Downstream Performance

At this point, we have an approach for preventing the loss of throughput. Now it's time to turn our attention downstream to the place where revenue is actually generated. Throughput for the whole supply chain is only unrealized potential until customer sales occur. We need a way to measure performance of downstream links. What kind of commitment should be expected of downstream links? In most cases, it seems to be, "If a real demand for end products exists, we will sell them and request replenishment." This is too vague to ensure the kind of performance we desire.

Certainly, retail stores are expected to generate high levels of sales, and they must be committed to making this happen. But we also expect stores to be able to determine sales expectations for new products. While we don't expect an accurate forecast in all cases, a relatively dependable estimate of "maximum" sales within replenishment time isn't unreasonable. Closer proximity to the market should enable sellers to be somewhat more realistic about customer desires and preferences than the other links in the supply chain.

The payoff for the whole supply chain depends on the downstream links' ability to move stock quickly enough to justify upstream investments in capacity and materials. Because we want to know how fast our inventory moves through the supply chain, we need to know how long an item takes to sell and how much that item costs. Additionally, we must be able to determine how much money is buried in inventory produced, but still not sold.

Inventory turns indicate how fast an investment in materials moves over time. It's possible to measure inventory turns for a family of items or a specific item, though few firms actually concern themselves with individual items. Another helpful piece of information for managing supply chain performance concerns breaking out the movement of a particular supplier's components in an end product through the chain. Remember, even for component providers, nothing is really sold until the end product is sold.

However, inventory turns are a measurement of the past, dividing average sales (or average cost of materials out of the sales) by average inventory level. If we want to decide quickly whether to invest in producing a particular product or where our sales efforts would best be directed, we need an effective picture of the current situation *now*.

It's possible to know what inventory is stuck between producers and stores, and the cost of materials that make up that inventory. But if we could also discern how long each completed item spends in inventory, we would be able to see blockages in flow clearly in near-real time. Inventory turns are an inadequate measurement to help us address these issues.

## Inventory Value Days

TOC provides an even more telling metric: inventory value days (IVD). It considers the investment in inventory to support supply chain's desired sales rate. It also enables identification of inventory clusters that may be sitting for an extended time and not supporting sales at all. Let's illustrate the IVD concept with an example.

Suppose the raw material cost of end item X is $2. The central warehouse inventory at the production site currently carries 250 units. One hundred of these were produced just two days ago. The other 150 units were produced five days ago. In addition, the distributor's warehouse (the link after the central warehouse) holds 400 units, of which 200 are 10 days out of production and 200 are 12 days out of production. Retail stores have another 400 units spread among various locations, of which 120 have been in the chain for 15 days and another 150 are 22 days post-production. The remaining 130 units are 75 days out of production.

The calculation of the IVD for that item is:

$$\$2 \times (100 \times 2 + 150 \times 5 + 200 \times 10 + 200 \times 12 + 120 \times 15 + 150 \times 22 + 130 \times 75) = 40,400 \text{ dollar-days.}$$

This is equivalent to holding \$40,400 worth of inventory for one day.

What is important to note in this example is that if the stores used extra efforts and sold the oldest 130 units of end item X then the IVD will go down to 20,900 dollar-days. That means those 130 units that are stuck in the system for a longer time have driven up the investment to quite a high level and it points clearly to the impact of the blockage in the sales of that item. If this information had been revealed 20 days earlier, then either it would have initiated sales efforts or maybe even have made a decision to discontinue the production of the item. If we decide to stop selling a particular item, we reduce the target inventory levels for that item to 0, thus stopping any further production. Practically, such a decision taken 20 days ago would have left the total stock of that item to be only 150 + 130 = 280 units (the 150 produced 22 days ago were completed before the assumed decision was made).

Of course, we do not know about the recent sales of that item. But because we still have stock from 75 days ago, it probably means that the stock level 75 days ago was much too high relative to the sales. Another explanation is that the 130 units are stuck at certain stores (they are really at the stores' level, not in the distributor's warehouse), while sales at other stores are adequate. If this explanation is true, then the appropriate action would be to pull back the stock from where it does not move well and transfer it to the other stores. Dynamic buffer management could have pointed out those stores holding too much stock, but a manager must make the decision to shift inventory to other stores. Also, because buffer management is not sensitive to the financial costs of the inventory, our inventory could contain a number of items whose stock is high. But focusing on hard decisions, such as stopping production of an item, requires information that is sensitive to the cost as well as to the quantity and time.

The IVD is often viewed incorrectly as the twin measurement to TVD. Actually it is something quite different, many times complementing the TVD. While TVD measures due date performance (see Chapter 4), IVD adds the perspective of the investment in inventory. Without going into the technical aspects of both measurements, let's agree that it is far easier to provide the TVD and/or throughput-loss assessment measurements than IVD. But the competitive advantage offered to the supply chain is large enough to justify including these measurements in the chain's performance measurement system.

# Conclusion

We believe that a supply chain employing the make-to-availability and the distribution concepts presented in this book can gain a significant competitive advantage. The appropriate starting point for implementing the concepts may be anywhere in the chain, but because they are so close to the market the distribution links are excellent choices. In many cases, the distribution links are collectively the largest and most powerful portion of the supply chain. Thus, implementing these concepts in distribution companies would impose pressure on suppliers to "do the right thing" by providing vendor-managed inventory using make-to-availability procedures. Such a win-win scheme, originating with the most powerful link in the supply chain, should attract effective attention.

It's important for the distribution links of the chain to reduce slow-moving inventory and to improve the availability of fast movers, because doing so improves the profitability of the whole chain. The replenishment of sales signals the remainder of the chain about the customer's taste. Of course, a distributor with superior response to customers is an asset in any supply chain, but without the active collaboration of the suppliers it would be difficult for the distribution chain to gain that level of competitive edge that could be a real breakthrough. The chains that adapt their logistics throughout the supply chain to the simple and effective methods we have presented will have the ability to gain a significant edge in the market.

The producer with superior performance in maintaining availability to the distributor promotes and enables superior sales support. Similarly, raw materials and parts suppliers must perform well by maintaining availability for the producers or else the producers' performance may suffer. But without the distribution chain ensuring excellent availability at the stores and active in identifying what goes well and what is slow, the overall performance will still be below the true potential. The fact that the chain is a series of links makes the results at one link dependent on its feeding link, and so on throughout the chain. A shortage anywhere in the chain may have far-reaching effects.

Our logistical concept for the ideal supply chain includes two main objectives: (1) We must have very fast replenishment times throughout the chain, and (2) We must also hold relatively low levels of stock at each link in the chain to allow for fast introduction of new products. In order to achieve these objectives, all links must ensure availability and work together to resolve any problems that arise. Each link must continue to improve (shorten) its replenishment time in order to reduce their required investment.

This logistical scheme has to be supported by the appropriate business practices. The farsighted rules of partnership lead to the idea of dividing the throughput generated by every sale to a customer to every participant in the sale. To tie the chain members together tightly we suggest payment to all links only after the end item is sold. The supply chain might need to take on another partner to manage the financial transactions.

Appropriate performance measurements should be in place to motivate the trust that is required for organizations to truly collaborate for the mutual gain. Therefore, the suppliers, producers, distributors, and the sellers should be evaluated based on their contribution to the chain's success in delivering the final product to the customer. In order to recognize deterioration in performance before it results in lost sales, we identified TOC measurements based on throughput (throughput loss and TVD) and inventory (IVD) to assess upstream and downstream links in the chain, respectively.

Developing and maintaining a win–win culture is at the heart of the approach we have outlined in this book. Is a win–win culture likely to lead to supply chain success? We think the answer is a resounding YES. We appreciate the idea of a win–win culture as a way to motivate chain members to work and collaborate together so that everyone in the chain truly wins. In times where customers' tastes change frequently, having win–win cultures may be the best way to promote supply chain success.

## Reference

Dettmer, H. William, *The Logical Thinking Process: A Systems Approach to Complex Problem Solving*. Milwaukee, WI: ASQ Quality Press, 2007.

# Epilogue

*"To be continued..."*

Writing a book on new knowledge that continues to develop is challenging. A book has a deadline. What happens when new ideas have evolved after the book reaches the reader? When knowledge is profound, it should remain valid for a long time into the future. But additions and clarifications will eventually be necessary if the material is to remain relevant.

The Theory of Constraints (TOC) is a continually evolving managerial approach. This book, concentrating on a new understanding of manufacturing and distribution, deals primarily with ideas that were developed during the past eight years.

Consequently, this book is intended as a point of reference for new knowledge that is likely to evolve. The technology to keep abreast of this evolution is the Internet. You, the reader, should know that we plan to add articles, simulations and case studies via designated websites to provide you with updated materials, as well as supporting the knowledge addressed in this book. Our aim is to make the transfer of evolving knowledge as efficient and effective as we can.

Additional material may be found and downloaded from the two websites indicated below. One of them is operated by Inherent Simplicity Ltd., a developer of state-of-the-art software for TOC applications in manufacturing and distribution. The other is maintained by Bill Dettmer, one of the authors of this book. Please check these sites periodically for updates and new developments:

1. Software: http://www.inherentsimplicity.com/warp-speed
2. Articles, and other supplementary materials: http://www.goalsys.com/supply_chain_at_warp_speed

# Index